P9-CQV-989

The House of Owls

Tony Angell

Foreword by Robert Michael Pyle

Yale UNIVERSITY PRESS/NEW HAVEN & LONDON

THE HOUSE OF

Owls

Yale University Press books may be purchased in quantity for educational, business,
or promotional use. For information, please e-mail sales.press@yale.edu
(U.S. office) or sales@yaleup.co.uk (U.K. office).

Designed by Mary Valencia.
Set in Bulmer MT type by Jo Ellen Ackerman.
Printed in the United States of America.

Library of Congress Cataloging-in-Publication Data
Angell, Tony
The house of owls / Tony Angell ; foreword by Robert Michael Pyle.
pages cm
Includes bibliographical references and index.
ISBN 978-0-300-20344-8 (alk. paper)
1. Owls—North America. 2. Angell, Tony. I. Title.
QL696.S8A538 2015
598.9'7097—dc23 2014035976

A catalogue record for this book is available from the British Library.

This paper meets the requirements of ANSI/NISO Z39.48-1992
(Permanence of Paper).

10 9 8 7 6 5 4 3 2 1

When you go owling

you don't need words

or warm

or anything but hope.

That's what Pa says.

The kind of hope

that flies

on silent wings

under a shining

Owl Moon.

—Jane Yolen, *Owl Moon*

Contents

Foreword

By Robert Michael Pyle

Owls. The very word has excited me since I was a lad. Unlike the author of the extraordinary book you hold in your hands, I never caught, raised, or lived with owls. I encountered Hooty the Owl in Thornton Burgess's *Bedtime Stories,* loved *The Owl and the Pussycat,* and scoured the local public library's shelves for books on owls, just as I did for otters, seashells, and butterflies. Later I read accounts of people who lived with owls, ending up many years later with Jonathan Maslow's *Owl Papers,* Max Terman's *Messages from an Owl,* and Bernd Heinrich's *One Man's Owl.* And when I read the Harry Potter books, I thought by far the best thing in them was Harry's owl Hedwig, and all the other mail-courier owls. It's only too bad that as a kid conservationist, as I fancied myself, I was way too early for Carl Hiaasen's *Hoot.*

As much as I loved reading about owls as a boy, I was even more eager to encounter them in flesh and feather. There wasn't much chance in my postwar subdivision in Colorado. But it wasn't long before I escaped the ordered grid and barren young yards, wandering off to the High Line Canal, an old irrigation ditch on the edge of the actual countryside. There, one enchanted day, I watched as a great horned owl burst from an old magpie nest in a cottonwood—and all of a sudden, owls had become *real.*

Since that thrilling moment, I can remember the first sighting of every species of owl I've come to know in the wild: The first northern spotted owl, on its nest on a low big tree bough in Sequoia National Park. The first hawk owl, crowning a black spruce in the boundless taiga along the Alaska Highway. The first saw-whet, fishing the shoreline of a little lake near Olympia. The first great gray, early in the morning, in the Blue Mountains of northeastern Oregon, even bigger than I'd dreamed. The first pueo, gliding toward me like a plane with a face through mist on the shoulder of Mauna Kea. And certainly the first snowy, which until then had been a merely mythic bird—never there for me where everyone said it had been only the day before, or the hour. And then there it was at last, squinting against a cutting windborne snow on a frozen beach north of New Haven, Connecticut.

Actually, it was during my three years in New Haven that owls came to have one of their deeper significances for me. The owls I speak of neither flew nor hooted. They were ornamental owls—stone, terra cotta, wood, copper, and so on—decorating the campus of Yale University as a frequently repeating symbol of wisdom, learning, and all of the scholarly and courageous attributes of their familiar, Pallas Athena. When I arrived at Yale in the fall of 1974 for postgraduate studies, I got off to a rather slow start thanks to my own intimidation and a little uncertainty about my thesis plans. Once I began to notice the frequency of the owl motif in the college gothic architecture of the university and its various colleges, each one a campus within the larger campus, I began collecting them. At first this was chiefly a displacement activity, giving a little structure to counterbalance (or distract me from) my lack of confidence in being there. It wasn't long before the intimidation faded and my dissertation research gelled, so the reason for the displacement activity was gone. But by then I was so much enjoying the hunt that I continued owl spotting for the whole three years.

By the time I marched in Yale's 275th graduation exercise, I had tallied around seventy-five "species" of owls among its hallowed halls and graven walls, such as the great copper owl weathervane atop Sterling Library, the roundels of owlish gargoyles encircling the Law School's crocketed towers, and various carved owlets on door panels, reredos, and moldings. In the absence of actual owls they served well as a distraction from the heavy academic work. (Many years later, during a joint reading at Redlands College in southern California, my fellow writers and I looked up from our texts repeatedly to watch both barn and great-horned owls cruising past the open doors. Had Yale been like that, perhaps I never would have finished my studies!) The elegant blend of art and natural history represented by this indulgent pursuit remains for me nearly as memorable as Professor Charles Remington's inspired lectures on evolution, watching Meryl Streep's student performances in several Drama School productions, and that first real-life snowy, out on the frigid shore of Long Island Sound.

In the forty years since, how very many owls have graced my days and nights! Downy baby great horneds and barns in the basalt rimrock of eastern Washington. Banshee-voiced tawnies in the Somerset countryside. A bouncing, big-eyed burrowing owl peeking over a road-crest at dawn, where I'd been sleeping on the opposite verge, during a very slow hitchhike across the western United States. Short-eareds cruising the Skagit Flats in company with northern harriers and red-tailed hawks and tens of thousands of snow geese. Particular pygmies and screeches in particular tree holes, faithfully present year after year. The barred owl of British Columbia whose shrill "who cooks for *y'allll*?" convinced a credulous young enthusiast that he had finally recorded Bigfoot. The still-life of a great horned owl, hunched in the crotch of a willow in a Wyoming winter landscape, that I mistook for a bobcat; and the one that cleaned out our hollow tree of its flying squirrels—one broad gray tail on the lawn every morning until they were gone—and then moved on. The barn owl that stooped at my cat, and the one I helped restore to its parents and siblings inside the gable of a very high barn, after rehab. The long-eareds, like small totems, lining the limbs of pines in a thick windbreak in the Columbia Basin, watched with Roger Tory Peterson on a bird convention field trip in 1971.

Another ornithological luminary along on that outing was Tony Angell. I had met Tony in Seattle in the 1960s, one of a number of keen and talented naturalists who made the city so exciting and inspiring for a college kid like me. Our friendship went from there, and will soon span half a century. A few years into it, when I worked for The Nature Conservancy, Tony was chairman of the Washington chapter. By this time, it was already apparent that he was one of the premier sculptors, pen-and-ink artists, and writers anywhere working with birds and mammals. Tony's love of corvids and of owls and other raptors led to a long series of marvelous books. Maybe it was inevitable that he would one day write a book in which he laid out his lifelong passion for owls and what they have taught him. When I learned that such a book was on its way, I rejoiced. Here, I thought, will be the book to cap all the owl books I've loved before. And so it is.

The House of Owls is, simply, a delight for a strigiphile like me. But it will also delight any birder or naturalist, and all those who care about the living world and its more remarkable manifestations. The heart of the book for me is the title chapter, "The House of Owls," which relates the personal saga of a period when Tony and his family lived intimately with one dynasty of screech owls who shared their home habitat. The next chapter, "About Owls," gives the basic facts to understand how owls work, and how they fit into the broader context of life, and the third chapter describes how they have been accommodated in human culture. The remaining three chapters consist of detailed verbal portraits of all nineteen species of North

American owls. The genius of this presentation lies not only in its comprehensiveness, and how we come away seeing the birds from all sides, but also in its shifting point of view. The first chapter is a deeply personal narrative that carries us not only further into the subject, but also further into the author than any of his earlier books have done. The other early chapters are largely factual and objective. And the comprehensive owl biographies are a masterful blend: each species is introduced through Tony's personal experience with it, and then its image is rounded out with fully researched, up-to-date information on its distinctive traits and lifeways. These accounts are nothing short of fascinating. The brilliant mix of personal and factual renders the whole compulsively readable.

There is one more category of readers for whom *The House of Owls* gives cause for huzzahs, and I count myself among them: Tony Angell fans. A handsome, warm, and imposing man of good cheer and rare intelligence, Tony made a big impression on me those several decades ago that has only grown since. I've eagerly anticipated each of his books, and never been disappointed—except in one selfish respect: I've always wanted even more of the man himself—his personal take, his lyric reflections on his subject—than the strictures of the books have allowed. Now, in *The House of Owls,* this is what we get, yet with no loss to the factual basis of the text. The art and the science are mutually reinforcing, as Nabokov (another fine artist and scientist in the same person) expressed when he asked: "Does there not exist a high ridge where the mountainside of scientific knowledge meets the opposite slope of artistic imagination?" That's just where we find ourselves in *The House of Owls*.

All this talk of artistic imagination brings us at last to the aspect of Tony's work that may be dearest to him: his graphic art. Tony is highly regarded as one of the foremost sculptors of birds and mammals, in stone, bronze, and other media. I've long thrilled to his otters, alcids, and other animals that one encounters in public places around the Pacific Northwest and beyond. But he is also immensely skilled in two dimensions. Many of his adherents acquire his books as much or more for the drawings they include as for their scientific and literary content. *The House of Owls* furnishes a beautiful blend of both. Interleaved among these illuminating pages you will find nearly a hundred exquisite drawings of owls in every posture, act, and attitude you can imagine. As Tony says, these drawings "are personal interpretations . . . based on direct and intimate observations." Clearly he values Nabokov's high ridge and is aiming right at it, for he intends the book to "build a bridge between those who want to observe owls and the subjects themselves, because I want to convey how we can feel about them as well as watch them." And he hopes through this approach to inspire readers so that "the observer of owls [will] become a student and steward of them as well." I believe they will, and I am certain that every lover of Tony Angell's art will exult in this new gallery. No other birds, by a long shot, have such expressive faces, and Tony captures these in his drawings with uncanny felicity and grace, showing us their true charisma, emotion, and range of personality.

I've always been a lover of owls. Now, a very long time and many miles away from Hooty the Owl, I feel at last as though I have a place to go to fully indulge this passion. Of course, the best place is *out there,* in the night, among the owls themselves. But when I can't do that—or maybe afterward, by the fireside—when I hanker to learn more about the birds, relive them through splendid portraits of words and ink, maybe plan the next outing, all this in the good company of my old friend Tony—I shall betake myself to *The House of Owls,* and walk in.

Preface

I dare say a flock of books on owls has been produced over the past several decades, and with good reason: owls are fascinating subjects. This book is distinguished from other such works by a personal narrative and a collection of illustrations that I have developed from a lifetime of living closely with many of the North American species.

The opportunity to have many of these birds residing in my home and yard for extended periods of time has provided me with information and experiences that are as emotional as they are scientific. This book is thus a record of the many years that my family and I have encountered these species: what we observed, what we felt as a result of these meetings, and what we have learned from owls.

Evolution has exquisitely designed owls for their lives as predators. They possess memories of place that are so keen they can maneuver expertly through the branches of trees in near total darkness. They are inquisitive, passionate, aggressive, deceptive, and at times quite valiant creatures. They experience pleasure and fear, and form inseparable pair bonds. As we humans make our impact felt on ecosystems and further pollute our planet, these birds are among the most vulnerable to the changes. The drawings and narratives here all grow directly from first-hand experiences with a number of owl species, but it is only by considering them in the context of the environmental conditions owls face that they become truly meaningful.

I begin by recounting the nearly quarter century my family and I lived in close company with western screech owls. This singular experience is presented as a chronicle of a full year in the lives of these owls and explores the growth of respect and attachment I developed for them. I follow this account with a discussion of how owls are different from other birds and explore their unique characteristics. To convey the powerful effect that owls have had on humankind, I provide a taste of the cultural history that features owls and a summary of why owls have been such an artistic focus in my life. The concluding chapters are an annotated look at owls found in North America from the standpoint of those that coexist with people, others that are somewhat specialized in their habitat requirements, and the ones that reside principally in wild and remote places.

I hope this book will build a bridge between those who want to observe owls and the subjects themselves, because I want to convey how we can feel about them as well as watch them. Our encounters with nature are frequently truncated and brief, and we are often inclined to take a glance rather than a thoughtful measure of what's out there. This narrative might help the observer of owls become a student and steward of them as well.

My illustrations are personal interpretations and by and large based on direct and intimate observations. Although my time and experiences with owls are unique, I would hope readers might find a model for how to write about and sketch birds. With perseverance and effort, they may wish to put a pencil to paper not only to record how they feel about what they see and hear of owls but, where words fall short of describing what they witness, to sketch them as well. Journal keepers who account for their days in the field might find the summaries of my experiences to be a template for their own. It has been a surprise for me, as I've developed my notes and quick sketches, how indelibly strong the experiences with the birds have remained. Owls have a significant place in our memory.

My approach has been to place the emphasis on the artist and naturalist's response to these remarkable birds. The personal tone reflects the enormous admiration I have for my subjects. By combining my moments of intimate contact with much of the scientific knowledge at hand, the reader's experience with these species is expanded—hopefully to a point where a decision we make regarding the condition and future of our environment will not be rendered without considering the welfare of the owl.

Birds in general have always held a unique spot in our minds—owls particularly so. A bird's power of flight will always leave us in awe, but owls also occupy the time when we are typically inactive, and which we know comparatively little about. Their presence in the night invites us to look, listen, and consider the full spectrum of the nocturnal ecosystems of which we are a part. To do so provides immeasurable rewards from personal engagement with the tangible, and the knowledge and creative opportunity that ensue.

Acknowledgments

Special thanks to the dedicated and knowledgeable scientists who compiled the accounts of North American owls for *The Birds of North America*. Their information has served as the basis for the life history summaries of the nineteen species included in this book, and may be consulted online at http://bna.birds.cornell.edu/bna. Alan Poole of the Cornell Laboratory of Ornithology, responsible for the range maps of North American owls from *The Birds of North America,* supportively provided me the use of these valuable materials.

Of particular help among the scientific community were professors Richard Canning and Fred Gehlbach, both experts in the field of owl scholarship, who consistently responded with requested information and support for my efforts. I owe my gratitude to Drs. Frank Richardson and Sievert Rowher, who, early on, provided access to the wealth of owl information at the University of Washington's Burke Museum.

My great thanks to Les Perhacs, Tom Jay, Jeff "Red Rascal" Day, Gretchen Daiber, Thomas Quinn, Fen Lansdowne, Don Eckelberry, Ivan Doig, Bert Bender, Marty Hill, Kevin Schaffer, Mike Hamilton, and Paul Bannick, all artists in their fields and friends who at one time or another have inspired me and provided me with the enthusiasm and support necessary to sustain my efforts on this work.

Very special thanks to Dr. Paul and Anne Ehrlich, scientists whose research and publications have significantly advanced the importance of understanding the complexities of nature and our place within it. Their energy, wisdom, and friendship over the years have been an inspiration.

Of particular importance has been my deep friendship and association with the artist and photographer Mary Randlett. For more than forty years her belief in and documentation of my work have been essential in maintaining my creative momentum.

Dr. Riki Ott, Ysbrand Browers, David Bennett, David Barker, Andy Haslen, scientists and artists from my days afield on the Copper River, thank you for the opportunity to be enriched by your knowledge and skills.

Photographer and technician Gregg Krogstad, thank you my friend for expediting the flood of drawings I needed documented, formatted, and forwarded.

A special thanks to my editor at Yale University Press, Jean E. Thomson Black, who believed that there was a place for this work in the vast collection of owl literature and patiently helped me fashion its form and purpose.

Tony Kavalok, thank you for sharing your story included here. And Noel Angell, I thank you for your early patience and sense of adventure that was part of first opening up our lives to those of the owls.

To my children, Gilia, Bryony, Gavia, and Larka, your genuine interest, help, and involvement throughout the years with all things owl is part of the heart of my story, and I thank you. Thanks also to Florence Angell, my mother, who permitted me the freedom to roam and collect in nature as a child, thus providing the foundation for what I've sought to say as an artist and naturalist.

Greatest appreciation and thanks goes to my wife Lee, whose keen editorial eye helped me navigate through the seas of information and technology to bring this collection of thoughts, observations, and art into focus. This ship would never have reached its destination without your energy, skill, and commitment.

The House of Owls

ONE

The House of Owls

Late in the summer of 1969, my wife and I moved to our first home, north of Seattle, Washington. This was an older house far from the city, in a location that had at one time been a northern destination for a Lake Washington ferry. The remoteness had preserved two streams that coursed through our neighborhood, both of which still hosted runs of salmon each fall. Kingfishers as well as green and great blue herons fished the waterways, and we could find footprints of long-tailed weasel, mink, otter, and coyote along the banks. It was something of an oasis, and even today some of this natural heritage remains intact.

By the end of the year we had settled into our home, and soon discovered what a severe winter storm can do to comb out the decayed limbs and snags of an older forest. When an intense storm hit, we got our first listen to the wind roaring through the boughs of the cedar, hemlock, and fir. The dead wood cracked and snapped. In this tempest, the ground shook when an ancient snag lost its footing and plunged to the floor of the woods.

That night, to get a feel for it all close at hand, I went out from our porch to the rain-swollen creek only a hundred feet away. The winds bore down and, as if to catch a breath for a moment, paused between the surges. At one of these intervals I heard something. It was one of those sounds that, while familiar, I couldn't consciously recognize. I cupped my ear to hear it again as another wave of wind raged through the trees, and nothing could be heard over the din. An interval of relative quiet and

1.1 A pair of western screech owls in a magnolia. When the young were about to fledge, this pair of owls routinely roosted below its thick canopy opposite the nesting box.

1.2 A pair of northern flickers. The abandoned cavity nest sites of the northern flicker are crucially important to western screech owls as places to lay their eggs and brood their young.

there it was again: a persistent, soft, and mellow sound. In spite of the night's fury, it was a western screech owl singing to advertise its territory, the same song I had listened to as a child more than twenty years before. Perhaps there was a nest here.

The storm passed, and the following morning I set out hopefully to search the woods for the bird's nesting site. The hunt was brief. A hemlock snag had been felled by the night's wind, and in a portion of its broken top a flicker had excavated a cavity that was split open. In the fractured nest, there were breast feathers from a western screech owl. This was certainly one of the nest sites the little owl had been declaiming about the night before, and perhaps the female had already been in it. This valuable piece of woodland real estate was now useless to the owls.

It was December, early enough that I thought to put up a nesting box for the birds, hoping they might breed close by. I set to work and hastily constructed a box from cedar fencing and an apple crate. What I lacked in precision I made up for with nails and an ample amount of wood. Although not a showpiece, it would go up in a hurry and not be easily storm-damaged unless an entire tree went down. The one aspect I was careful about was the diameter of the entry hole near the top of the box face beneath the slanted roof, which measured about three inches (seven and a half centimeters). I knew the owl would be partial to an opening that fit its body and matched that of the flicker's nest.

The box measured sixteen inches (forty centimeters) high, with the cavity inside a little over a foot from the bottom of the entry. The four vertical sides were twelve inches wide (thirty centimeters). I added a sprinkling of wood shavings to the interior to provide comfort to a brooding owl. I hauled the fifteen-pound box up twenty feet and permanently secured it to the massive trunk of the cedar next to our bedroom window. It faced southeast, and I christened it "the Fortress." We just had to wait and see whether the owls might occupy it.

January. The New Year arrived; the box had been in place for two weeks. Nearly every night I heard bits of owl calls bubbling up from different locations in the woods. Nonmigratory, these owls remain in their territories throughout the year, so I assumed the owl was making the rounds and defining its boundaries. This behavior continued through the month. It was not until February that the owl made it clear the Fortress was an acceptable nesting site.

February. By four in the afternoon on cloudy days it was already dark, and when I stepped out for a quick stroll along the creek I walked into a stream of owl singing that was clearly up-tempo. The accelerated song seemed to radiate a very determined declaration that the male was in residence and had property to show. It was a rhythmic and sustained *"Whoo-whoo-whoo-whoo-whoo . . . whoowhoowhoowhoo-whoowhoo."* The song accelerated in its rate, and was best described with the oft-used simile "like a bouncing ball." The intervals between the whoos decreased until I heard them as a roll of continuous sound.

It was easy to discover the owl, because he was but a few steps from our porch, perched on the roof of the nesting box. As I stood on the path below, the owl reared upright and puffed up his bright white throat patch and changed his muted appearance into a demonstrative display. Thrusting his body forward, he pitched his musical proclamation into the recesses of the darkening forest. Repeatedly he sang his short song, each time turning in a different direction, then pausing for a few seconds to catch his breath and probably listen for a reply. When the owl turned my way, the

1.3 Western screech owl vocalization sequence. The male begins advertising his territory early in the new year. Standing atop a prominent perch, the owl puffs up his throat, leans forward, and produces his signature call. He then quickly looks this way and that, listening for a response. The sequence is then repeated over and over again hundreds of times through the night.

white feathers of his expanded throat seemed to be illuminated and put the focus directly on the performer like a spotlight. "Here I am," the bird declared. "Look at me!" There was cause for celebration in our house, because here was a sign that we were to have a new neighborhood family move in.

A heavy, wet snow had fallen, but the chill and breaking branches seemed to make no difference to the owl. When the light faded, the owl went to work. His calls increased in intensity and frequency. At two o'clock one morning, with my bedroom window wide open, I was roused by the bird's calls. His bouncing-ball song was incessant. After a few minutes I started to count each sequence. In a span of two hours, at a rate of eight calls a minute, he had produced nearly a thousand individual runs of his song. In fact, this was only the beginning, as the owl continued to sing throughout the night and into daybreak.

One had to be in awe of the owl's energetic commitment to courtship. Ahead, of course, was the big test of convincing a mate that he was a partner fit to supply adequate food for her and their brood. Would a female accept the Fortress as a suitable site to raise their young? She would make the choice. I couldn't help but draw some comparisons with rituals of human romance. After nearly a month of effort by the owl, would this all be in vain?

As I sat on my porch two nights later, I heard a reply to the singer above me. The call was lower in pitch and softer, but unmistakably a bouncing-ball call. The two birds continued to sing back and forth through the night, all in the general area of the nesting box. Eventually the songs grew so close together that they became a duet.

March. By the middle of March the nesting box had met the approval of the female, and the two owls were defending the perimeter of their territory. They had altered the advertising call with an occasional bark, and directed their exclamations across our yard toward the road parallel to our property. From the other side of the road came an equally vigorous series of calls from another pair of owls. It seemed to be a border dispute akin to two human neighbors arguing over where the property lines were drawn. On the other hand, perhaps the box I had placed was such a hot item that more than one pair was competing for it. I would never know, since a second night of competitive scolding did not occur.

As I watched one late afternoon, the male owl landed on the box with a house rat in its beak, entered the box, and reemerged without his catch. I judged that the female might have been in the box already, unless the rat was left for her as a further demonstration of his prowess as a hunter and provider. Gifts like this were an integral part of the western screech owl's courtship, along with the mutual preening I frequently saw.

Mutual preening was often a prelude to mating. The pair would move along a perch and converge, the female leaning forward to expose the nape of her neck to the male. The male responded with eyes shut, to nuzzle down into his mate's soft plumage.

1.4 Western screech owls allopreening. Preening each other is an important part of the pair-bonding process in owls.

After the male groomed the female, she extended herself flat over her perch, and the male would in turn flutter up to suspend himself above her, steadying himself with a grip on her nape and flapping his wings. The coition was over in only a few seconds. Mating was a good deal more than just ritual, and occurred more than once during the course of the night.

April. The buds on the magnolia tree threatened to burst, obstructing the view to the Fortress. Although the owls could fly directly up into the cavity, the foliage of the adjacent tree provided some shade from the direct sunlight that came as spring advanced. It was the second week of April, and only the male called occasionally from afar. I heard him bark in protest at something unseen in the shrouded woods. He was on watch, and the female was in the box beginning to lay her clutch.

I confirmed that the female owl was on eggs by climbing to the box. Keeping my intrusion as limited as possible, I maneuvered my hand into the box. Its diameter restricted my entry to as far as my wrist, which was far enough as it elicited a sound like chattering teeth as the owl snapped her beak at me. Although not impenetrable, this box was a fortress that raccoons and opossums would find impossible to enter very deeply. Whereas inspecting the eggs was a temptation, further exploration could well have encouraged abandonment and discouraged subsequent nesting in years to come.

The forest canopy had thickened appreciably. Varied thrushes sang fervently from the tops of the tallest Douglas firs, and the robins hit their first notes of the dawn chorus of passerines. Our daughters were as excited as we were to learn that an owl family seemed to be under way. We watched as the male made several visits to the box each night with food parcels for the brooding female. Mice, crayfish, a single songbird—all were delivered to the hungry mate.

The nights were quiet, save for the *yip-yip* of the neighborhood coyote. I reminded myself that it was reasonable for the owls to keep a low profile while the eggs were being brooded, and later as the vulnerable youngsters were maturing. I considered myself the expectant father pacing outside the delivery room, and was about to climb to the box to inquire when I noticed a familiar form amid the snarl in an adjacent Douglas fir. The male had moved his day roost even closer to the nest to maintain his watch. His presence assured me that the female was safe inside and the hatching was imminent.

Steller's jays woke me one morning. There's a unique pitch I believe they reserve for predators they particularly disdain, owls among them. Going outside I found the jays and triangulated from their postures to discover the male roosting inconspicuously. The jays spent nearly half an hour scolding before they moved off. The male was still there when I checked late in the afternoon, and his reluctance to flush from the cover suggested a strong attachment to this place, which allowed a direct line of sight to the box.

The female now called to the male at all hours with her faint whinny begging solicitation. If the male didn't respond she increased the intensity of her calls and gave them more frequently. I watched her mate move in response directly above the box to roost in the magnolia, but he made no attempt to hunt. Still she cried, incidentally alerting a chorus of small birds, which began to mob the male.

When mobbed, the resident male and female paid little attention to the smaller birds. With a tree to his back the owl appeared to be sleeping, as black-capped and chestnut-backed chickadees led a charge flying back and forth over the owl. A cluster of bush tits took a top-row perch to offer high-pitched, wispy calls. Some of them appeared naive about the owl's location, looking out in the wrong direction, still scolding. The mobbing served an important purpose to these woodland species, because it introduced the owl as a possible predator to those birds that had little or no experience with it. I suspected this was also an opportunity for some of the male

1.5 Male western screech owl scolded by Steller's jays. Jays would often discover the male screech owl at his roost and, with a scolding pitch and intensity that seemed reserved only for the most threatening predators, alert the entire avian community to the owl's presence.

mobbers to demonstrate their fitness by confronting the owl. And while the clamor did little in this situation to move the owl out of the neighborhood, it did identify his location and that of the nest. An incident like this would impress the smaller birds, and they would remember it as a place to avoid.

Early May. In the first week of May, the female suddenly made an appearance in the entry to the nesting box. It had been more than three weeks since I had seen her, and I guessed that the eggs were hatched. Seeing me, she remained in place, feathers puffed up to fill the dark entrance with her body. When she closed her eyes to mere slits, her form blended into the matching background of the surrounding wooden

1.6 Songbirds mobbing a male western screech owl. With his mate in the nesting box, the male owl roosted closer to the nest and more in the open, where a host of local birds would gather to scold him.

box and tree bark, and she was nearly invisible. A few days later I heard soft chirping calls accompany the female's constant whinny reminders to bring home the bacon. The young owls were in the nest. Our daughters came from the house to listen, and we watched the male arrive overhead, dimly silhouetted and holding food in his beak for the family. We celebrated with whispered "*Hoorays!*"

By the middle of the month the male owl was delivering food to the box every hour or so through the night. Over this time he shifted his roosts to all sides of the box and occasionally changed locations during the day. Although rarely in the same spot, he now remained perched within a radius of twenty-five feet of his brood.

1.7 Female western screech owl bumping male into flight. When the brooding female was particularly hungry, she occasionally bolted from the cavity to hasten the male's hunting activity by bumping him off his perch.

Charting the course of the western screech's emergence as a family in our woods had instilled a concern for the birds' welfare that settled in with all the Angells. Every day, arriving home from school, our daughters would ask about the owls and check on the nest box. Friends would be introduced to our resident owls, and short summaries of their manner and habits would be shared. To a degree our well-being was measured by how the owls were doing, and our understanding of how the remaining wild community functioned here was predicated on the birds' presence. That our experience had fostered a recognition of our capacity for altruism was indisputable.

The male is the sole provider for the family until the last few weeks before the young ones fledge. I worried about the possibility that something would befall the male and the entire family would starve. My imagination would occasionally get out of hand to the point that I considered trapping mice or getting a colony of them to raise in case such a calamity should occur. Finding road-killed owls as I did from time to time throughout the year, I was reminded of the mortality of these birds, even though the far greater percentage of such fatalities is among the young owlets, not adults. In any case, I would routinely look for the roosting male each morning to assure myself he was safely on watch and prepared to hunt through the night.

Early one afternoon the calls from the young and the female in the box grew loud enough to be heard from inside our house. I went outside in time to see the female bolt from the box, wings aflutter, to approach the male on his roost. He didn't budge and appeared indifferent to her solicitation. He actually appeared to be feigning sleep. She was having none of this apathy and took a step forward to soundly chest-bump him. Knocked off balance, he stood upright and wide-eyed, and she bumped him again, sufficient to knock him off the end of the limb and into flight to the woods. It was several hours before sundown, but I was sure the owl's night of hunting had begun.

Last Week of May. It had been three weeks since the owlets began hatching. I was certain by their calls there were at least two hatchlings. On a very warm midday the female occupied the nest entry panting visibly. I imagined that the box interior could grow uncomfortably warm. As I watched, she still gripped the bottom of the nest entry as she sprawled down to rest her head on the box perch below. Unmoving, she looked like an enormous gray banana slug as she dozed there, probably seeking relief from the muggy interior of the box.

The female spent more and more time outside the box, often blocking the direct sunlight that might heat up the interior even more. Her presence on the perch on the face of the box discouraged the inquisitive eastern gray squirrels as well. Likewise, keeping the cavity plugged with her body kept egg-laying flies and biting insects from entering the nest.

The owls tolerated our gardening activities, and one or the other of our daughters would mow up and down the overgrown lawn with little attention given by the birds. Our Siberian husky, Quinn, was active and noisy, and he too seemed to be considered nonthreatening.

Both adults were out hunting now, as the rambunctious brood continued to grow. In the early evening I saw the pair, beaks bristling with carpenter ants, delivering food two to three times an hour. There must have been a nearby nest of ants to provide such largess. Less frequent deliveries of larger food offerings were made into the box by both birds. Exiting adults often emerged carrying debris from partially

1.8 Female western screech owl dozing at front of nest box.
As the interior of the box got warmer with growing young and rising spring temperatures, the female owl would lean out of the entry and doze with her head on the perch.

eaten prey and castings, and they would drop it at some distance. Letting waste accumulate near the nest would be a magnet for a hungry raccoon or opossum.

On a very hot afternoon, the male appeared at the nest with his breast heavily saturated with water. It appeared that he had entered the cavity this way, perhaps attempting to cool off the nestlings with a shower of water.

June. The pair of adult birds now roosted outside the box. No food-begging by the female, as the owls had settled into the upper canopy of the magnolia. From my desk I made out their blunt outlines against the transparent green above them. Their positions appeared to be strategic, not random, because they were perched on opposite sides facing the nest, where they had a direct view of it should any threat appear.

1.9 Adult screech owls roosting strategically.
As the young owls matured, the female moved out of the nest cavity
to roost with the male on either side of the box.

These were long afternoons, and I periodically checked on the birds to see if they had changed their sentinel positions. It was around this time that I observed the male owl hunting down a Swainson's thrush that was singing in the nearby forest. Hearing the songbird, the owl snapped out of its indifferent roosting posture and flew off in the direction of the thrush's song. A minute or so later the owl returned, holding the thrush in one foot, and flew directly into the nesting box.

With the longer days I had the opportunity to follow the owls as they began their early evening hunting forays. Initially, the male launched into a particular tunnel through the branches that took him to creekside. Like a driver knowing the road well with its twists and turns, the bird flew at what seemed high speed, anticipating the branches in his path and easily moving this way or that to avoid them. I usually

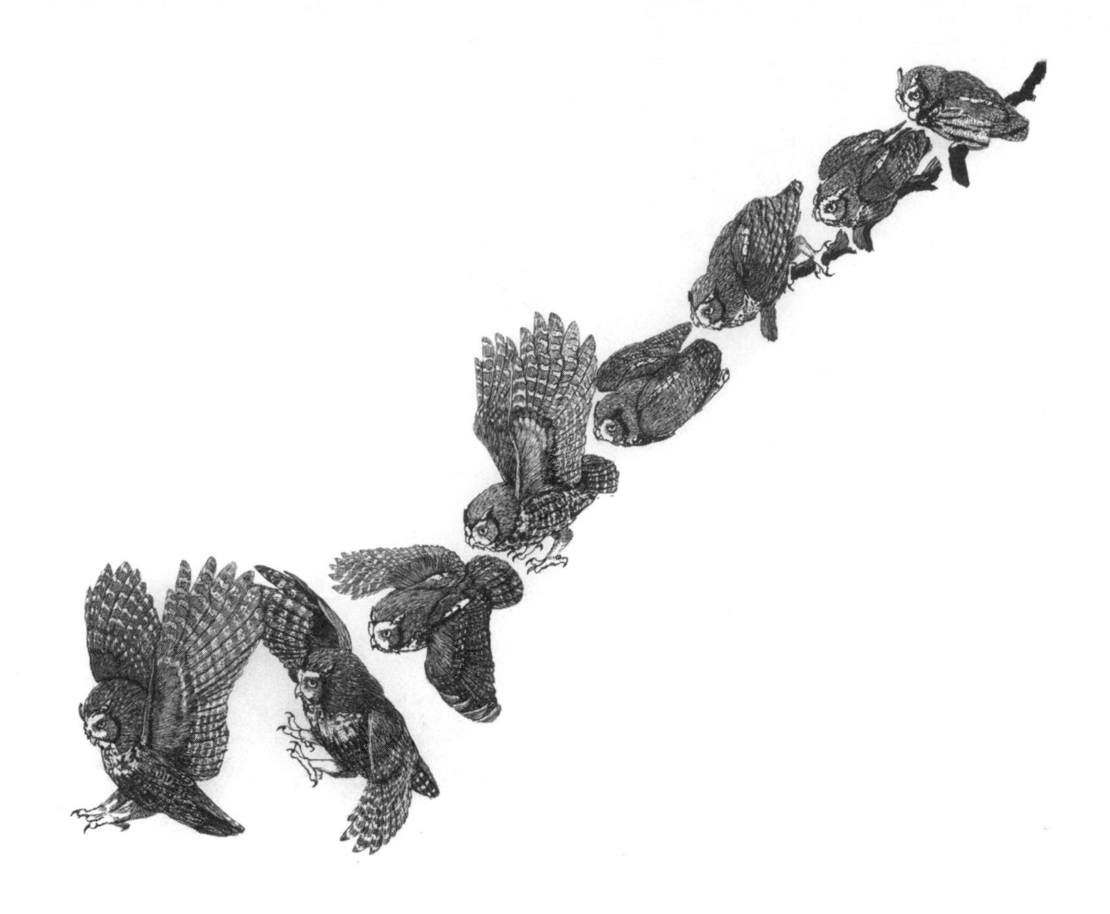

1.10 Sequence of owl descending to strike prey.
When the owl sights a prey target nearby on the ground, it descends rapidly, accelerates with a few wing beats, and then pulls up just before striking to extend its legs and taloned toes to grab its quarry.

found him spending his first few minutes surveying the sandy banks at the water's edge, the site of many crayfish captures. If the action was slow, he flew straight away to the next perch along his route, another location above a creek segment where crayfish could be caught scavenging in the shallows. I followed at a respectful distance, but eventually he took to the adjacent woodlands, and the groundcover of salal and Oregon grape was too dense for me to follow.

Retrieving from roadside an injured or deceased owl that had collided with a car has given me the opportunity to appreciate the remarkably large feet this smaller species has. Being zygodactylous, they have two toes that are positioned forward and two backward. This arrangement is advantageous because it allows the birds to

1.11 Western screech owl about to strike a mouse.
The wide spread of the owl's big taloned toes contributes to its success as a formidable predator of small rodents.

hold their prey securely, particularly some of the more slippery aquatic species they capture. The fledglings used their strong feet for gripping perches during their initial, rather flightless launch from the nest, and for climbing back up from the ground to their elevated perches.

When I caught sight of one of the adults making a prey capture, I gained some sense of its technique. The western screech may first be alerted to the presence of prey by a sound. It will then get a sightline on its quarry and drop off the perch before gathering speed with the pumping of its wings. Just before contact, the owl extends its legs and feet in front of its body with toes wide open to strike the prey. For aquatic prey, the owl would take an entirely different approach and wade into shallow water to snatch a crayfish or retrieve caddis fly larvae. The sure-footed owl also snatched up small rainbow trout that fed in the shallows near the creek's surface.

One morning, beginning the second week of June, some twenty-eight days after they hatched, two feathered faces appeared at the cavity entry. The youngsters had climbed up the interior of the box to take a first look around at the world beyond. The nest portal was something of a bottleneck for the inquisitive owlets, but pushing out they both bobbed their heads from side to side to register a sharper image of what they were likely seeing for the first time. As I stood below and opposite them, they seemed fixed on my form, and I was certainly transfixed by them. I was struck by this mutual interest we showed in one another, free of any fear, an uncomplicated connection between two very different species. At the moment, it seemed that there was no predisposed or instinctual avoidance that separated us, just a simple intent to take one another's measure.

Our daughters joined me for this innocent encounter with these peaceful arboreal creatures. We marveled at their energy and beauty, and even laughed at their super-animated movements. It was strange, since we were all touched with the same feeling that we wanted to get even closer to the birds—know them even more—but that would be a mistake. In spite of the temptation, it was important to respect the boundaries that set us apart, to keep them wild and prepared to take care of themselves in a dangerous world.

The days that followed the first appearance of the young always included one or the other or both of the owlets occupying the entry to the Fortress. They watched their parents roosting above them, followed the routes of passing insects, and confronted a scolding Steller's jay before retiring to the safety of the box's interior. If Quinn was lounging on the lawn below, they studied him as well. It was one revelation after another for the young birds, which were now getting so large only one at a time could look out from the box. This was a troubling arrangement for the owlet left inside. On more than one occasion as I watched, the owlet perched in the entry would get yanked back inside by its nest mate, which then climbed up for its turn at a look outside. The pull of a life beyond the nest was becoming irresistible.

The Fortress now sounded like an electronic keyboard amped up to play a medley of whinnies, chirps, hoots, and barks. The birds worked on their growing vocabulary of begging, protests, and scolds. Our family felt the tension building in the owl family as the young sought to separate from the confines of the nest.

1.12 Owlets viewing the world from nest box entry.
With their first views of a world outside their nesting cavity, the owlets appear to be interested in anything that moves.

1.13 Steller's jay scolding an owlet at the box.
An owlet's otherwise inquisitive nature is short lived
in the face of a Steller's jay.

A little over a week from the time they first glimpsed the larger world, the youngsters exited the Fortress. They had grown to more than three-quarters of their adult size, and while they were not yet prepared to do any flying, they capably fluttered, jumped, and then walked out on to the nearby magnolia branches. The big feet of the small owls served them well. They fastened a strong grip on whatever they encountered in their initial leaps to the tree branches, and with the aid of their beaks and fluttering of stubby wings, they climbed into a perched position.

Near the middle of June, I was out first thing in the morning to watch the owl family that had been roosting together outside the box for three days. The rotund adults and their offspring had tucked themselves into the foliage overhead. In the course of one day, a parent left and returned with a crayfish, and the fledglings

1.14 Adult female screech owl protecting an owlet.
When jays came to scold a youngster at the box entry, one of the adults would take up duty on the perch and shield the owlet from the clamor.

clamored in anticipation of a meal. The adult sectioned the catch, and the cartilaginous tail was given to a youngster that bolted it down. Its nest mate's attempt to grab a portion was in vain. The following morning I found a casting beneath the tree that included a portion of the indigestible carapace of the crayfish. It was still slick, well-lubricated with mucous for regurgitation. I opened it up and there inside, neatly assembled like some collapsible miniature plastic toy, was the complete exoskeleton of the crayfish tail. Judging by the long time it took the owlet to swallow this mouthful, I think it was nearly the outside limit of what its gizzard could accommodate.

On the fourth day since the fledging of the owlets, I was working outside at my studio carving when a racket began to build from the side of our property where the

1.15 Owlet swallowing a crayfish tail. After the adult had sectioned the crayfish, the young owl would bolt down the tail, carapace and all.

owls were roosting. The scolding noise quickly erupted to a cawing frenzy as crows were recruited from all points to join the melee. Within minutes of my arrival, the number in the mob grew from half a dozen to nearly thirty birds, an amount and level of frenzy sufficient to qualify as a murder of crows.

It was all going on beyond my reach, and I stood helplessly and watched. The marauders were so intent on directing their wrath at the owls that they hardly noticed me when I grabbed a sizable broken branch from the ground and tossed it ineffectually in their direction. Unless you have spent time amid a body of overwrought crows, you have no idea what an intense din they can create.

The entire episode lasted no more than a few minutes, but possessing such paternal feelings for the owls and knowing their vulnerability in such an onslaught, it seemed much longer. One of the crows flew up the path in my direction as it sought to come at the owls from below. As it did so a brown-gray boomerang shape dropped from above,

1.16 Female strikes a marauding crow.
To protect her brood and against what appeared overwhelming odds, the female screech owl dropped from her perch and struck one of the crows in the clamorous mob.

striking the crow squarely in the back of the head. The blow caused black feathers to erupt from the bird's nape, and it momentarily wavered toward the ground before recovering enough to fly off awkwardly over our housetop. A swirl of crow plumage drifted about me as I watched the female owl loop back up into the family's roosting site. With the owl's attack on the crow, the emotional climate was instantly altered. The bully birds, now at bay, suddenly and silently retreated.

The odds had seemed overwhelmingly in favor of the crows, and I wondered if my presence had made a difference in the outcome. I wanted to claim some credit, but the adults had demonstrated remarkable resolve in holding and defending their ground, and I hoped they would have succeeded with or without me. What was certain, however, was that the young owls' initial innocence had been replaced with a very important memory and wariness of crows. It was all part of the rapid and intense early education of the screech owl—what to avoid and what to fight—with more to come.

Later that same afternoon as I walked along our woodland path, I discovered that one of the owlets had somehow grounded itself. With its lemon yellow eyes aglow, it looked up through the groundcover of oxalis. It was certain the adults overhead were well aware of the youngster's location and would feed it, so that was not my concern. A greater worry was that it might fall prey to the coyotes and raccoons that patrolled this area, should it fail to get off the ground before evening. I considered whether to leave it to its own devices or to pick it up and place it in a nearby tree. I chose the latter, in spite of the intrusion it might have on the natural order of things. My commitment to the bird's well-being was more active than that of a scientist who might have waited it out and accepted the consequences, come what may.

Having rightly or wrongly given myself the role of godfather to the family of owls, I scooped up the fluff ball that was the owlet at this stage. It fit comfortably into my open palm before edging up to perch on my finger. My daughters soon joined me, and they took in the subtle beauty of this bird they could now study so close at hand. Here we were again, two different species that stared at one another with fascination. I couldn't help but make something of a comparison of my daughters at this stage and this young owl hardly more than five weeks old. Innocent, inquisitive, and still dependent on the vigilance of their parents, both had much to learn for the full potential of their lives to be realized. The owl served my daughters as a lively doorway for learning. Just the sight of the owl close at hand opened up the imagination, the process of inquiry, and the initial knowledge that leads to wisdom.

1.17 Grounded owlet in oxalis. A young owl that ended up on the ground following the incident with the murder of crows was found amid a thick cover of oxalis.

1.18 Western screech owlet being held.
When finding an owl on the ground, or any bird for that matter, it is tempting to give the bird a hand up to a more secure perch.

The twins and I discussed the importance of getting the owlet off the ground and considered a route back to its family. A compromise was reached by placing the bird alongside a rhododendron that offered an opportunity for the bird to jump into its lower branches. From there it could work its way into the adjoining tree boughs. After a quarter hour, this was exactly what it did. As we watched, it made a foot-over-foot beak-assisted fluttering ascent to the lower boughs of the cedar and rejoined its nest mate.

End of June. By the end of June a period of fluctuating weather had beset us. A sunny day would be followed by days of rain and cool weather. On the hot days both the adult birds and the young sought the shade of the magnolia and perched well back in the thickest portions of the foliage. Even in cooler retreats the birds panted rapidly to relieve the buildup of heat in their bodies. In such weather, I would occasionally find the male owl with the feathers around his stomach, flanks, and lower breast soaked from a recent cooling dip in a side pool along the creek.

1.19 Owlet climbing back into a cedar tree.
Even though fledgling western screech owlets do not fly in the first few days from the nest, they can climb by using their beak and feet and fluttering with their wings.

When storms dumped water, I became concerned, but expected the same umbrella of magnolia leaves that lessened the impact of the heat would also protect the birds from the driving rain. I checked, but found only one owlet in the retreat. An adult owl's plumage, being as soft as it is, does not shed the rain easily, but the still-downy owlet was a dry sponge waiting to be saturated. Still struggling for full

1.20 Female screech owl sheltering an owlet in a downpour.
The rainstorm could have compromised the well-being of this recently fledged owlet,
so its mother provided some protection under her wing.

flight, the wet youngster could be grounded, but moreover, with the soaked feathers losing their insulation value, the bird could become chilled and die of hypothermia. I walked the immediate woods searching the groundcover, looking for the owl. The wind picked up, and the rain increased as I headed back to the house. By chance I looked up and caught a glimpse of one of the adults perched in a vine maple. I thought it odd to find it in the relative open under such circumstances, with one wing extended out over the branch it stood on. When I stepped back for a better look, all was revealed. Beneath her wing and hard up against her flank, I could see the face of the missing owlet. The mother owl was sheltering her fledgling from the deluge.

July and August: owl diaspora. Ten days had passed since the youngsters fledged. Much stronger and capable of confident flight, they abandoned their natal

1.21 Young owls observe adults capturing prey.
Watching the older birds as they hunt is an important part
of the education of a fledgling.

site and started to follow their parents each evening on their foraging pursuits. I was not surprised when I found them at dusk along the creek, where the male owl would often begin his hunt. The youngsters, perched a short distance away, witnessed his flight to the edge of the water where he captured a crayfish.

Such was part of the young birds' education. They learned what to eat by what was brought to them at the nest, and they learned the fundamentals of where and how to catch it by observing a capture in process. It wasn't all a matter of instinct, and this time of year the watershed was teeming with insects, small birds, and one litter of mice and rats after another that the youngsters could fatten up on as they honed their predatory skills.

Moving out of the old nesting location was a necessity. Mobbing birds, some bothersome and others dangerous, knew the site well and came looking for the owl family. Clustering together at a predictable site would make the owls more vulnerable to predators, including the local Cooper's hawks that nested annually in our watershed. The bigger owls that moved through these woods were also threats to these smaller owls. Dispersed farther north and following the creek, the family remained together for another month or six weeks and called back and forth in the evenings as the summer progressed toward fall. *"Whooo . . . Whooo . . . Whooo,"* the exchange seemed to begin with a question: "Are . . . you . . . there?" A pause and then an answer: "I . . . am . . . here."

So well had the owls committed to concealment it was rare that I discovered them at roost. For the most part they remained silent or at least toned down their calls to the point that I would rarely hear them. The Fortress looked ragged and lifeless, having lost its vitality with their departure. Abandoned human homes have a way of looking that way too. No longer a home, only an empty and hollowed structure remained.

September and October. The silence of these months of autumn suggested that the owls had disappeared from our woods. It was strange, however: even though there was no sign of their presence, I was confident that they remained. I couldn't consciously account for this certainty, but I felt it. In any case, it was good that they stayed out of sight and carried on their separate lives uninterrupted by my curiosity. The young birds had yet to master a body of owl knowledge that I could not even imagine. I knew that, as the owlets expanded their range to find territories of their own, it would put them into unfamiliar places and expose them to unforeseen dangers. It would be remarkable if even one of the two fledglings was able to make it through its first year of life.

With the owls gone, I climbed to the Fortress and with a scoop filled a grocery bag with the debris that had accumulated. For a number of reasons the cleanout would be a good investment. It made it more likely that, if the owls chose to return to the box, it would be partially cleared of parasites that might survive into the following year to infect an adult or young owl. The material that accumulated over the several months of owl occupation also provided a pretty good accounting of what the birds had been feeding the family and hinted at some of the ecological dynamics of this riparian location.

Scoop after scoop of dry nest contents filled the bag, until I got nearly three quarters of a pound of it from all but the far corners of the box. A single, unhatched egg, which I later measured and weighed, came up immediately as I spread out the remains on a sheet of cardboard. The organic material wafted an odor like duff one would pick up by the handful from the forest floor. Sorting through it, I found a mixture of evidence that supported the description of this species as a very opportunistic feeder. The compacted hair of rodents, contained in the owl's castings that remained in the box, made up the bulk of the leftovers. I credited the adults with catching a dozen young house rats and an equal number of deer mice. The indigestible portions of a wide range of insects appeared as well. The owls had a great appetite for the larvae of the northern caddis fly they snagged out of the creek. The protective encasements of the fly were throughout the nest material. Entire castings of the hardened parts of our big carpenter ants were also abundant, along with the discarded wings and legs of yellow jackets, bumblebees, assorted coleopteran beetles, crickets, sow bugs, and mourning cloak and swallow-tailed butterflies. Among the more aquatic species consumed was the crayfish that, judging from the amount of their indigestible parts found, was a special favorite. There were a few preserved tail parts from small trout. The case for our owls having any serious impact on our local songbirds was very weak. Only the olive-brown tail feathers of the Swainson's thrush were present, along with flank feathers from a spotted towhee.

Survey completed, I swept the material back into the bag, but kept the unhatched egg for further study. I sprinkled the bag's contents around the old cedar and about the wild rhododendron where my daughters and I had placed the wayward fledgling owl many weeks before. It seemed a proper and nurturing reward to the plant that had given up a branch or two for an owlet climbing to rejoin its family.

Winter solstice. From my studio the forest appeared as weary of the weather as I was. Snow had started to fall, but I brightened up a bit when I thought of this shortest day of the year in light of it being the date of the birthdays of my mother, my wife, and my twin daughters. We all retired early; better to sleep than resent the climate. I passed our open bedroom window, and a sound outside brought me to a halt before

1.22 Eastern gray squirrel takes up residence in an owl house.
In the months after the owl family had vacated the nesting box, other animals from squirrels to honey bees sought to occupy it.

I consciously identified what I had heard. I leaned out into the dark and could only hear the creek thrashing about in its bed. Then—there it was again, and again and again. The little screech owl was singing, advertising his brave presence in the face of an impending storm. The following night, walking by a casement window in our dining room, I looked out to see the owl eyeing me as if we had met before. The owls had returned.

Afterward. The owls continued to use the Fortress every year from 1970 to 1994. Over that time it appeared that at least five different pairs of owls used the box for nesting and occasional roosting. All together they fledged about fifty young, and judging from my annual analysis of their diet, they made a significant impact on the population of small rodents living along the creek. They also reduced the numbers of ants and termites that damaged our home.

Based on nearly a quarter century of living compatibly with this species, I can say with confidence that these are remarkably adaptable owls, even if they were modestly tended to by us. I have no doubt that the birds learned, similar to many corvids, to recognize the individual faces and manner of my family members, and allowed us to approach them and get much closer than they would have let strangers. And perhaps the most profound parts of this period were the experiences that living with owls afforded my family and me. They routinely invited us into their natural world, where in their inimitable way they informed and taught us.

Cycles in nature are the rule and part of the dynamics of changing ecosystems. How the occupancy of the owls in the Fortress came to an end after twenty-five years was nevertheless surprising. One year in our woods I began to hear the booming calls of the barred owl. It had been reported here in western Washington since the late middle part of the twentieth century, having arrived in the Pacific Northwest by moving across Canada from the East Coast. Big, aggressive, and predatory on smaller owls, it came into the watershed during the last year that the screech owls fledged a family. I suspect that the barreds hunted down and captured the youngsters, and perhaps the adults as well. The smaller owls, having no experience with this threatening relative, had yet to evolve a strategy to avoid it. The discovery of barred owl castings containing the feathers of the western screech owl provided evidence of their role in the disappearance of our owl family.

It is interesting that the eastern screech owl, having coevolved with the barred owls in the East, coexists with the bigger owl in the hardwood forests there. Coevolution played a role in the smaller owls' development of instinctual and learned behaviors that made them less likely to be captured by barreds. There may be other factors that protect the eastern screech from predation that could otherwise lead to its extirpation. Knowing these factors could help us anticipate the needs of our western species. Perhaps time, and a lot of it, will allow the same evolutionary process to occur in the behavior of our western screech owls.

I was angry and sad, and felt helpless in the face of the loss of "our owls," as we came to refer to them. They had provided my family with such intellectual and emotional sustenance. But with some thought, it would give little satisfaction to shoot the messenger, either literally or figuratively. The barred owl was following opportunity in its movement across the continent, as humans changed the landscape and made it more suited to its feeding and breeding requirements. We widened the road west for the barred owl and paved it, too.

1.23 Screech owl in winter looking into dining room.
The screech owls remained in our woods throughout the year and would sometimes be seen outside looking in at us.

To whatever degree we can, we must preserve and exercise stewardship of the existing contiguous forests and pristine riparian habitats. We know what smaller owls require to thrive. When we eliminate retreats for their breeding and roosting, and satisfactory conditions for their prey, the birds suffer.

Here at my home, I still listen in the early winter for the lively sound of screech owls. The woods are silent, except for the creek that continues its indifferent rush to thread through the trees down to the lake. Not long ago, however, I was savoring the taste of salt and kelp that blew up from the bay below our studio on Lopez Island. It was a quiet night early in the year, and I heard the unmistakable song of the western screech owl floating delicately from the forest behind me. With its *"Whoo-whoo-whoo—whoowhoowhoowhoo"* it exclaimed with authority "I am here!" again and again. I whistled back "I'm here too," and I'm hopeful.

TWO

About Owls

Owls have a long evolutionary path, with fossil records dating from the Miocene Epoch, some 23 to 25 million years ago. Over time, they have diversified to where today there may be, depending on how one classifies them, as many as 217 different species of owls in the world, and the list is still growing. In the order of owls, Strigiformes, there are two families. Tytonidae represents barn owls, and Strigidae includes all other owl species.

Owls are found on every continent except Antarctica. Exploiting a range of habitats from bleak and frozen tundra to steamy jungles, they vary in scale from the sparrow-sized elf owl to the Blakiston's fishing owl, which is as large as some eagles. At one time owls were even bigger. Their evolutionary history includes a fossil of an immense barn owl recovered in the Caribbean. Active less than thirty thousand years ago and three times the size of today's barn owls, it was capable, like a harpy eagle, of taking giant sloths from the jungle treetops.

We don't ordinarily consider owls to be either common or diverse, compared with the gulls we might encounter along our coasts. It's certainly true they are not as abundant in numbers, nor are they typically concentrated where they live, but around the world there are more than twice as many species of owls as gulls. Because of their secretive and reclusive habits and being active mostly when we humans are not typically out and about, they simply conduct their lives with most of us hardly knowing they are there.

2.1 Size disparity among owls. Owls have an unusual size range, as seen in the difference between the elf owl, which is not much larger than a song sparrow, and the Blakiston's fishing owl, which weighs as much as a small eagle.

As the avian family tree is arranged, owls are not closely related to hawks, eagles, and falcons. But through the process of convergent evolution, these families have developed similar physical and behavioral traits to be efficient predatory birds at different times in the twenty-four-hour daily cycle. In general, owls can do pretty much the same thing in periods of limited light that hawks, eagles, and falcons can do during the day.

The success of owls as predatory species rests in part on their ability to carry out much of their lives in relative darkness, avoiding competition with other avian predators that occupy full daylight. The generalization that owls operate in total darkness is not entirely accurate, because it is estimated that less than half of the world's owls are exclusively nocturnal. Even then, there are various degrees of illumination in which the birds choose to be active. Complete darkness may be found in the bottom of a cave, but on the earth's surface total darkness really doesn't occur. Some owls routinely forage during the diurnal periods from sunrise to sunset, such as the short-eared, pygmy, snowy, and hawk owls, and although they may select for different prey, they share their habitats with hawks, eagles, and falcons. The twilight period shortly after sunset is also favored by the short-eared along with snowy and great gray owls. Owls such as the barred, spotted, barn, saw-whet, and boreal are most active in darkness tempered by starlight and moonlight.

Worldwide, if feeding and breeding conditions are right, there are few habitat types these species cannot occupy. In North America, snowy owls are well established in some of the farthest reaches of the Arctic, occasionally sharing the space with the ubiquitous short-eared owls that, along with their close relative the long-eared owl farther south, can be found across North America. In the taiga forests of the far north begin the ranges of the great gray, hawk, and boreal owls. Here, in and about the Arctic Circle, the great horned owl can be found in a few numbers as well, and continues its population spread across a variety of habitats to cover the entire continent. Sharing space in portions of these northern forests is the saw-whet owl, a smaller but close relative of the boreal owl. Northern spotted owls hold on to a tenuous existence in the few remaining old-growth forests of the Cascade Mountain range southward along the West Coast and into the southwestern United States, where a subspecies maintains a small population. The aggressive and opportunistic barred owl has extended its numbers from the eastern and southeastern forests across the northern portions of the continent to occupy the western woodlands and become a serious competitor with its close relative, the spotted owl. Eastern screech owls are established in forests westward to midcontinent, while the western screech owl maintains its presence throughout the northern and western coastal forests, down into the Cascade and Rocky Mountains, to the Southwest and Mexico. In the West, the diminutive northern pygmy owl occupies some of the same range as the western screech owl, as does the flammulated owl.

Ferruginous pygmy and elf owls raise their young in stands of saguaro cactus and also divvy up the available nest sites of the dry oak forests of the American Southwest with whiskered owls. In the arid open lands of the West and portions of Florida, ground-nesting burrowing owls share the habitat with mammals. And finally, perhaps the most adaptive owl of all, the barn owl, has established a presence across the more temperate portions of the continent, exploiting both our agrarian and urban communities for the inevitable small mammals that accompany human enterprises. A good argument can be made that these species have coevolved with humans throughout the world.

Owls not only occupy a wide range of habitats, but also enjoy an unmatched diversity in size, which opens up a wider range of niches in which to settle. At one extreme, the diminutive elf owl of the American Southwest functions effectively as a largely insectivorous owl. About the size of an average man's thumb, this tiniest of all owls weighs less than a medium-sized hen's egg, about two ounces (55 grams). On the heavyweight side, we have the enormous Blakiston's fishing owl of the Asian Arctic. Weighing more than a good-sized bald eagle at ten pounds (4.5 kilograms), these well-insulated birds can capture and subdue salmon weighing nearly as much as themselves. Although the lightweight elf owl needs little insulation in its desert habitat, the massive fishing owl is blanketed in layers of feathers, allowing it to maintain its body temperature in the severest of Siberian winters.

The apparent size of an owl can be very misleading, however, inasmuch as the layering of feathers required by many owls for their life in the elements makes them appear much larger and heavier than they actually are. I have often spoken to audiences about owls while sharing some of my drawings of them. Showing a life-sized rendering of a great gray owl, I would ask the assembled group how much this enormous bird weighed. Of course, I set up the audience for a bit of a surprise by using the term "enormous," but it is a big bird. People's responses often compare the owl's apparent size with an animal well-known to them. "As much as my elkhound," someone will say—"about thirty pounds." It isn't unusual to have a respondent compare the owl to someone in his or her family. "I had my baby on the scale this morning and he now weighs as much as that owl; around eighteen pounds." There is always

2.2 Great horned owl colliding with a windshield.
Collisions with automobiles are a frequent cause of owl deaths.

collective disbelief when the audience learns that the average great gray owl weighs no more than three and a half pounds, or about one and a half kilograms.

Even those among us who know the outdoors can be fooled by the appearance of a big owl. A friend of mine, a seasoned hunter and fisherman, was returning from a trip one evening when a great horned owl flew up from the road and collided head-on with his windshield, shattering it. "Biggest owl I've ever seen. Must have weighed fifty pounds," he said. When I told him that four pounds is about as big as this species gets, his animated response was an emphatic "B.S.!"

The effect of the impact on his windshield became more plausible to my friend when we discussed the fact that the speed of the owl's flight combined with the velocity of his oncoming car was analogous to a four-pound cannonball striking the glass at eighty miles an hour.

Although some owls sit and wait for prey to appear, others have different strategies, including stalking or striking their quarry. Their vision, however, is fundamental to their success, and all owls can see perfectly well in daylight. Fewer than half of the world's two-hundred-plus species are truly nocturnal. These are the species of owls whose auditory sense takes over when their vision cannot discern the prey image sufficiently to strike it.

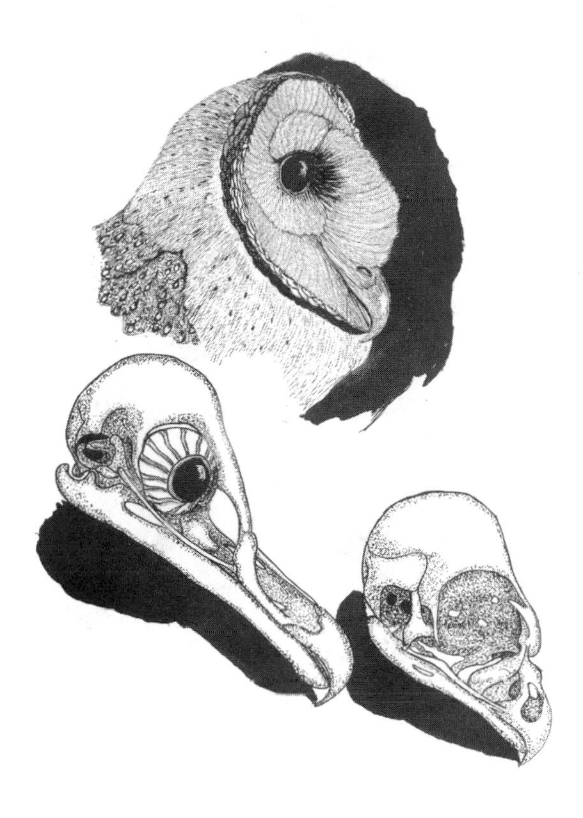

2.3 Barn owl's light-collecting face and elegant skull compared with a hawk owl's skull. With unique facial feathering and eyes extending from its skull, the barn owl can collect and focus all available light to see an object. Its extended beak is particularly suited for reaching out and subduing rodents, whereas the beak of the hawk owl is more truncated.

The eyes, however, are the most important sense organ to most owls. They provide an owl with exceptional visual acuity, and with fused sclerotic rings the eye is lengthened, allowing the entry of more light. Some of the eagle owls have eyes that are larger than a human's, and their extension outside the skull also provides the birds with a binocular view of the world. Having the cartilage support the eye rather than it being encased in bone lightens the weight on the front end of the bird. This allows the owl to avoid being top-heavy and keeps its aerodynamic balance intact.

Even though the owl's frontal field of vision is less than what we enjoy, their ability to quickly swivel their heads around 270 degrees provides them with an almost instant capacity to respond to sound or movement and scan the landscape for prey.

2.4 Saw-whet owl's eyes, one contracted and one dilated fully.
To adjust to light levels, owls can contract or dilate their pupils well beyond what the human eye can do.

Our average 180-degree head-turning ability requires us to shift our entire body to get a better look at what's behind us.

The retina of the more nocturnal owls is several times more sensitive in perceiving low light levels than the human eye. As one would expect, these species hunt in response to shape and movement, and possess a greater density of light-sensitive rod cells than the color-sensitive cone cells. Furthermore, the elongated eye allows for an extended cornea and lens, increasing the amount of light that enters. The owl's iris can close down the pupil to a mere pinprick when light conditions are intense, and conversely in darkness it pulls back or dilates expansively, allowing every possible amount of a light to enter the eye. Studies have demonstrated that when looking at the same image an owl, possessing this larger pupil, will see it more than two and a half times brighter than a human does.

The feathered facial disks of all owls concentrate the available light, giving the bird the capacity to better resolve images under low-light conditions. These disks surround each eye before meeting in a crested shape above the bird's beak and forehead. Light is directed to the eyes from all angles, and although it is done for entirely different purposes, the light concentration is not unlike the devotee of the perfect tan who, at poolside, holds metal reflecting panels to either side of his or her face to maximize the tanning effects of the available light.

Owls vary in their visual capacity to operate effectively in nocturnal circumstances. Some, like the northern pygmy owl, with its smaller facial disk, and the hawk owl, with a body like a forest hawk, are by physical form and habit better equipped to hunt during the day than at night. Although the motivation of a hungry owl has an effect on its success in catching quarry, it is also speculated that owls that are largely nocturnal have developed a mental map of their surroundings that allows them to know and anticipate, without precisely seeing, what obstacles they face when flying through or hunting in familiar territory. Inside their hippocampus they hold a three-dimensional memory map of their home ground. In marginal light they can employ direct visual realization, but at lower light levels they can rely on memory as well.

I learned to appreciate this phenomenon of remembered space when a storm took out all the lights in our home. Determined to get to the kitchen, I easily made my way through the labyrinth of furniture, despite the darkness. My memory of where things were located in this familiar territory allowed me to move with some speed and avoid a collision. Likewise, with a visual memory of a familiar space stored, an owl is assisted in pursuing its prey in darkness.

Considering the squeaks and squeals that characterize the sound repertoire of small mammals, one can also expect that owls are more sensitive to higher frequencies of sound than we are. They are also capable of picking up the intensity of a sound that is well below our threshold of hearing.

The owl's facial disk is a very fine filigree of firm feathers. Although they collect and focus light, they also allow sound to filter through and reach the bird's auditory canals. When an owl hears an animal sound, it will turn to face it, and the disk of feathers with its border of a stiff ruff will collect and focus the sound as we might use a parabolic sound collector to gather and record something heard. For owls that hunt during the day, the sound will alert them to the presence of prey, and the sight of the quarry will be what the bird relies on to launch a strike. Some owls, however, are equipped with openings or apertures for hearing that are asymmetrically located

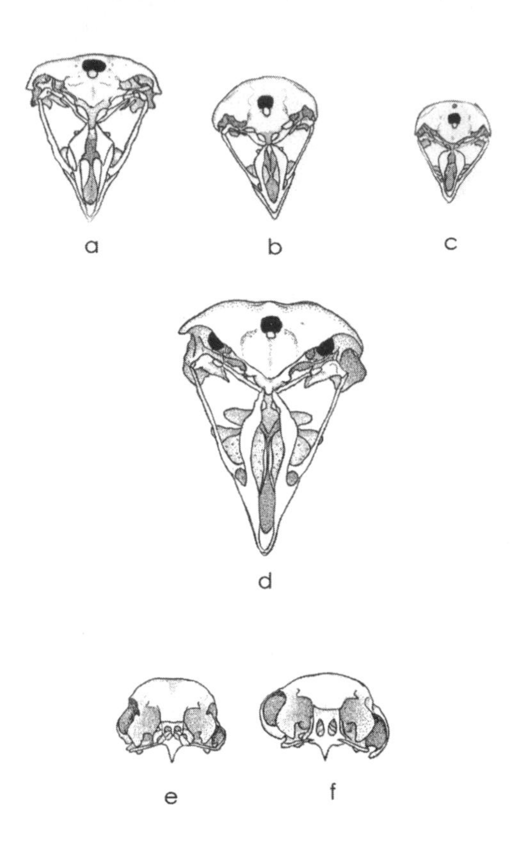

2.5 Owl skulls with symmetrical and asymmetrical auditory openings. The ear openings on the skulls of the diurnal-hunting hawk owl (a), burrowing owl (b), and pygmy owl (c) are symmetrical in their placement. The auditory openings of the great gray owl (d) are slightly asymmetrical, assisting it in detecting prey beneath snow cover, even though it is diurnal in its hunting habits and discovers prey by sight. The skulls of the saw-whet (e) and boreal owls (f) show dramatic asymmetry in the location of their auditory canal openings, which aids them in detection when hearing prey is more important than seeing it.

on their skulls, with one higher than the other. Even if the sound comes from behind a barn owl, its facial sound-gathering disk is also equipped with an ear flap beneath the feathered surface that can assist in directing the sound of prey activity. Should a barn owl hear the rustling of an animal coming from its right, it turns to confront the sound so both ears receive the sound simultaneously. Now facing the prey, the barn owl can determine if it is above or below its line of sight, as its left ear is higher than its right. If the sound is below this line, the sound will be louder in the right ear.

2.6 Great gray owl penetrating snow cover to capture prey.
Using its extraordinary sense of hearing, the great gray owl can detect and locate a small mammal beneath a blanket of snow and will plunge through an icy cover to seize its quarry.

The great gray and boreal owls can capture prey concealed beneath a thick layer of snow by positioning the head so the sound of the quarry enters the ears with equal intensity. By doing so, they know the location of what they seek on both the vertical and horizontal planes. With feet extended, they plunge through the snow to catch a vole prowling in its tunnel beneath frozen cover. The great gray can hit the snow hard enough to break an icy crust an inch thick and reach down into the cover more than a foot.

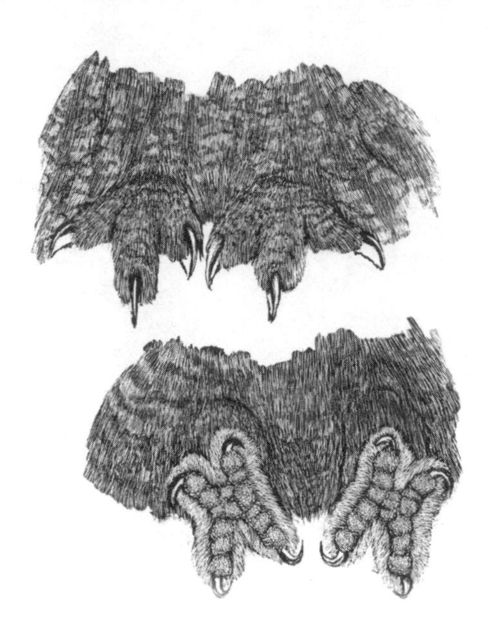

2.7 Great gray owl's toes, topside and bottom.
Although the great gray owl's toes are thickly feathered over their tops and edges for insulation, their undersides are uncovered and touch-sensitive, allowing the owl to feel and position its grip on what it might catch.

Even in flight, the great gray can make fine adjustments to follow the sound of a concealed moving animal. The soft plumage of body and wings dampens the sound of the bird's movement, allowing an unimpeded auditory connection to the sound of its quarry. Given the importance of sound detection to the owl's hunting success, it's no surprise that the hearing center of the owl's brain, the medulla, is very complex and possesses three times as many neurons as a comparable part of a crow's big brain.

It is one thing to rely on its unique ability to detect the sound of prey and to plunge through the snow to catch a vole sight unseen, and something else to feel the prey and snag it with its feet. The undersides of the heavily feathered toes of the great gray owl are bare skin, with touch-sensitive nubbiness, allowing the bird to latch on to a scurrying body it can't actually see. Our fingertips are, of course, loaded with touch receptors, and, thinking of the owl reaching below the snow cover to grip its prey, I'm reminded of being in the shower with shampoo-filled eyes and dropping the bar of soap. The touch receptors take over as I blindly reach this way and that, trying to grip a familiar shape. Using the sensitive undersides of its taloned toes,

2.8 Great horned owl positioning prey for swallowing. The bristles extending from both sides of the owl's beak assist it in feeling the body of the animal it is subduing and properly positioning its catch for swallowing whole.

the owl finds the familiar shape day after day. The featherless toes of a barn owl have bristle-like extensions that, similar to a cat's whiskers, assist in locating prey.

The facial bristles that surround the mouth of the owl also provide important information to the bird when faced with a struggling animal. These specialized feathers allow the owl to position the beak for strategic bites required to dispatch a thrashing squirrel or rat without exposing the eyes to the possibility of injury.

To remain light for flight, owls don't have teeth to chew their food, but they do crush it with their beaks. After a bit of softening up, the owl's meal is swallowed whole or in large portions, leaving the digestive process to separate waste from sustenance. As it closes in to bite the quarry, the bristles or filoplumes about its beak also assist the farsighted owl in lining up the body of the mouse, vole, or squirrel so that the entry into its mouth is typically head first, with the prey's legs pressed tight against its body for a more streamlined passage into the bird's throat and esophagus.

Watch a pair of owls strengthen their bond through mutual preening, and again you appreciate their sense of touch. With eyes closed (a protection, should either one object to the action), the male feels the feathered surface of his mate's nape and nestles into the plumage to administer what I imagine is the equivalent of my spouse scratching my back in places I would otherwise be unable to reach. It looks pretty clear that some pleasure is involved, and the endorphins are flowing in the female as she moves forward into the pressure to encourage its continuation. A barred owl I named Buttons lived with me for nearly twelve years and was fond of sidling over my way on his perches, extending his head and inviting me to push down through the tracks of feathers to lightly rub the skin surface on the top of his forehead and the back of his neck. There seemed to be no end to the pleasure as far as the owl was concerned, but fifteen minutes was my limit.

Holding the body of a great gray owl is similar to holding a big down pillow with a fresh sweet potato in the middle of it. This is an owl with a wingspan and body length nearly as large as those of a small eagle weighing eight pounds (3.6 kilograms), although the owl itself weighs less than half that. Like most of our North American owls, nearly silent flight is part of the great gray's hunting strategy. By examining the outer edge of one of its primary feathers, a partial explanation of its capacity for stealth can be made. In the middle of a ten-inch primary feather are more than 320 loose, eyelash-like barbs extending along its outer edge. These soft enhancements dampen any sounds that might be made as the wings of the flying bird cut through the air. This reduction in sound allows the bird to give full attention to detecting sounds its concealed quarry might make.

As with other birds, the great gray must keep its body temperature within a critical range, averaging around 40.5 degrees Celsius (105 degrees Fahrenheit). Encased in a suit of feathered insulation, the bird has less of a problem maintaining its warmth than it does cooling off in warmer weather. Its feathered head is as large as a small child's, but its skull and neck are encased in a layer of feathers 3 inches (7.5 centimeters) deep in some places. The tracks of feathers beneath its wings and along its flanks and chest are long and supple, and when elevated they trap warm air between them, giving the bird an outer insulating barrier against winter temperatures that routinely reach well below 32 degrees Fahrenheit (zero degrees Celsius). Underneath the contour feathers of the chest is yet another layer of silky down, and below

2.9 Snowy owl in flight.

this, adjacent to the skin, is a final insulation covering of very dense, downy stubble more like fur than feathers. Nearly immune from the effects of the cold, the bird can turn its attention to the greater challenge of finding sustenance to maintain its strength in habitats where other predatory birds are absent.

Likewise, the snowy owl requires a shield of insulating feathers to stay in its Arctic range during winter. Its plumage is somewhat denser than the great gray's, and although it is ideal as a covering against the far north's freezing conditions it can be a problem if the owls make their way southward during periodic irruptions. When wintering in more temperate portions of the continent, snowy owls can be exposed to elevated temperatures that may be life threatening. Birds lack the sweat glands that allow mammals to reduce the heat buildup in their bodies. An owl equipped for a colder climate and exposed to rising temperatures needs other strategies to escape or release the heat. For the spotted owl, its multilayered old-growth forest habitat provides various microclimates that the bird can use to remain comfortable through both the coldest and the warmest seasons.

Along Puget Sound where I live, snowy owls making their periodic visits will sometimes encounter unusually warm winter days. On more than one occasion I've watched the owls, perched out on drift logs and on the banks of sloughs, trying to cool off in the face of rising temperatures. Partially opening their wings helps by permitting greater air circulation to wick off the heat, and sometimes they pant like racing dogs. Mouths open, they flutter the gular pouch of the throat, allowing the air to wick off some of the body heat from the naked surface of the interior skin. On a few occasions I've seen the big birds flop out on the ground with wings open, looking like a sack of spilled laundry. There they hunker down on the cool sand or mud substrate along the shore for relief.

Given the effects that warming weather can have on the physical well-being of some species of owls, one can easily imagine what a calamitous effect global climate change could have on them. Among other factors, internationally agreed upon reductions of carbon emissions must be reached to assist in the stabilization of rising global temperatures. Failure to do at least this will hasten the decline and eventual extinction of species that cannot tolerate increasing temperatures—owls among them. It is sad beyond words to imagine that generations will not witness the evocative manner and stark, silent beauty of these species.

Along with the insulating feathering that distinguishes their order, owls also have layers of fat beneath their skin that can serve the same purpose. Unlike many waterfowl with a fat layer that is uniformly attached to the inner layer of the birds' skin along flanks and chest, great horned owls I have examined have patches of fat that are symmetrically positioned over specific locations on their bodies. Along the lower abdomen and across the lower back (areas that are exposed to the chill of lower temperatures), patches of fat are positioned, appearing to serve as additional insulation as well as a stored energy resource for the owl. The body of one male great horned owl I studied weighed 3.5 pounds (1.6 kilograms) and, of that total, fully a quarter pound (110 grams) was stored fat. By having uniform-shaped clusters of insulating and energy-providing fat placed in equal amounts over the surface of its body, the owl never lost the aerodynamic balance required of an avian predator.

The owl, like all birds, has special physical equipment that it brings to the task of living in a particular habitat and exploiting a unique niche.

Like us, owls have binocular vision, but because their eyes are fixed in their sockets, the field of sight is not as wide as a human's. They more than compensate for this, however, by the capacity to swivel their heads 270 degrees, nearly 100

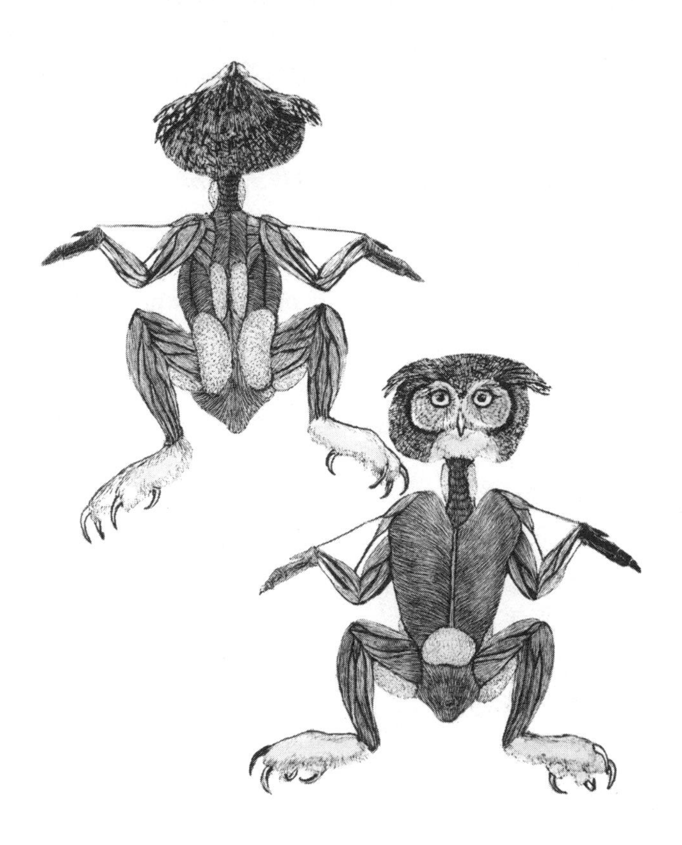

2.10 Great horned owl, dorsal (back) and ventral (belly) body views showing fat deposits. Fat is stored on the body of the great horned owl in a manner that maintains the bird's aerodynamic balance.

degrees farther than we can. Such movement is facilitated by the owl having fourteen vertebrae in their necks, twice as many as we have. Furthermore, the jugular veins are arranged in such a way that turning their heads to this extent does not disrupt the supply of blood to their brains.

Standing up on their toes, owls are hardly in the wading bird category, but compared with many other orders of birds they are rather long-legged. Barn owls can thrust their upper leg and extended foot into stands of grasses around outbuildings and in fields to grab small mammals in a manner resembling the long-legged northern harrier. The great gray owl, combining the weight of its body with the thrust of its long legs and feet, can reach voles a foot and a half (45 centimeters) beneath the snow. The leggy burrowing owls, while not quite up to roadrunner standards, are capable runners as they chase desert reptiles, mammals, and insects along the ground. True to their name, these birds also use their relatively long legs and stout toes to further excavate or expand the ground burrows dug by the animals with whom they share habitat.

Although there are subtle differences among species, the owl's four toes are typically short and stout, the better to hold and subdue prey. When clutching an animal it has caught or when perched, the outer front toe will swivel to face the rear, allowing it to place two toes forward and two backward. Once the owl has caught something or is comfortably perched, a locking mechanism in the foot allows it to maintain its grip without contracting the muscles.

The movement of owl populations is rather unpredictable, but it can be generally described in several ways. In the northern portions of North America, some species not equipped to remain in their territories through the winter will migrate year after year to more southerly locations where a food base is available to them. Some owls might also make a winter move from higher mountain elevations to the lowlands for the same reasons. There are also irruptions of owls. These occasional mass movements of birds into regions they do not regularly occupy are less predictable. Although irruptions of boreal, hawk, and great gray owls do occur from time to time, the most renowned certainly is that of the snowy owl. Theories abound why these movements occur, but current thinking suggests that an abundance of lemmings and voles in the arctic has resulted in a bumper crop of young snowy owls. With so many snowies competing for food, the young are forced to disperse farther southward in the search for sustenance. It's also fair to say that resident adults would not tolerate competition, and this intolerance may accelerate the initial movements of the owl's southward journey. Although not usually in the same years, significant numbers of snowy owls reach both the northeastern and western coasts roughly every four to six years. Some circulate into the midwestern portions of the continent as well.

In western Washington, their stunning and unusual presence is such a jolt that television sets are often abandoned on weekends in favor of visiting a local beach access where owls have been sighted. Here we can witness something far more spectacular than one more familiar episode of football, as our somber gray-brown winter neighborhoods and shorelines are lit up by the birds' big bright profiles. People are exploring the mysteries of nature rather than discussing a controversial call in the ballgame. During the winter of 2013–2014, owls roamed from their Arctic nest sites thousands of miles along the East Coast as far as Florida and Bermuda; imaginative naturalists among us are still talking about it. Whatever forces precipitated the

2.11 Young girl in view of snowy owls. When in the presence of wintering snowy owls, it's important to keep one's distance and allow the birds to conserve their energy and hunt undistracted.

movement, we can be confident that the birds possess an inherent capacity to wander in their winter food search and return to their natal habitat the following spring.

Migration also occurs in owls, although documentation of these movements is more difficult to monitor given the nocturnal habits of the species. Each fall, populations of flammulated owls from the northwestern reaches of their range migrate southward as far as Mexico. Saw-whet owls also have a southerly migration along the eastern portions of the continent. Somewhat distinct from a migration, northern pygmy owls descend each fall from their higher mountain elevations into the lowlands where temperatures are more moderate and prey are readily available.

Dispersing families of owls also results in unusual movement of these birds. A young saw-whet owl from the forested hillsides of Vancouver Island in British Columbia was recovered several provinces to the east, having flown through and

2.12 Flicker at nest cavity entrance.
Small owls generally depend on woodpeckers, and flickers in particular, to create cavities where they can nest.

around the formidable Canadian Rockies and into taiga-covered land. A long-eared owl, also banded in Canada, made a flight of more than 2,000 miles (3,200 kilometers) to Oaxaca, Mexico, where it was collected. From the standpoint of population expansion, the barred owl is a species on the move. Establishing its numbers westward across Canada and down along the northwestern coast over the past half century, its movements are best described as pioneering. As we shall see, these aggressive predators have displaced spotted, screech, and saw-whet owls.

Owls don't build their own nests, although the burrowing owl is capable of excavating or improving a mammal's burrow to suit its breeding requirements. Owls will move into a suitable cavity or a nest constructed by another large bird, often a crow or a hawk. A cleft, shallow cave, or cliff that occurs naturally is also suitable for

2.13 Barn owl with rat. A single barn owl, living only a short span of a few years, will catch and consume thousands of rodent pests, particularly rats and gophers.

some owl species. Barn owls have a proclivity for finding acceptable nesting surfaces in the abandoned or rarely used structures of humankind.

Without the flicker and other medium to large woodpeckers, the boreal, northern pygmy, eastern and western screech, saw-whet, elf, whiskered, and ferruginous owls would be hard pressed to find suitable roosting and nesting cavities. The nests of the magpie, common crow, and raven, and the red-tailed, broad-winged, red-shouldered, and Cooper's hawk supply long-eared, barred, and great horned owls with protected enclosures to incubate eggs and brood young. If older-growth forests didn't have snags to be excavated or hollowed out by the elements, the spotted, great gray, and hawk owls could not find proper protected enclosures to raise their families. To nest, the barn owl exploits deserted outbuildings, ledges along rafters in warehouses, and bell towers, and will take to nest boxes when available. Without these retreats to raise their young, our rodent-consuming barn owls would be far fewer.

If you spend any extended period of time with another animal, it is human nature to begin to notice and describe traits that we seem to have in common. I have no doubt that crows, ravens, and their kin with their sophisticated avian brains possess a range of emotions, some of which are similar to our own. I'm also certain that, to some degree, owls do too.

Using human terminology to describe another animal's behavior can be questionable. To overdo it would be to anthropomorphize. On the other hand, I feel that some of my observations derived from close association with these birds can best be described in terms that suggest behavior analogous to our own.

Can owls get angry? Judging from their behavior, I think they can get pretty close to it. I'm not just thinking of an emotional state that territorial owls might feel when their home ground is trespassed upon by another pair of birds, or what rival male owls might feel toward one another when seeking the attentions of the same female. Certainly the adrenaline is flowing in these circumstances, but there are more subtle conditions that have suggested an owl in a fit of pique. It's not unusual to watch our wintering short-eared owls respond aggressively to the rough-legged hawks and northern harriers hunting the same salt marshes. On one occasion I followed a short-eared owl that had just snatched a vole from a patch of grasses lining the edge of a slough. As it rose up and headed for a drift log to consume its catch, a northern harrier veered undetected in its direction, did a half-barrel roll, and snatched the vole from the owl's grasp. The owl responded with a loud and raspy *yaaaak,* suggesting an exclamation someone might make if accosted by a purse-snatcher. The food pirate's escape was so fast that pursuit was pointless, and the owl began to circle up over the mud flats. After a few minutes, the bird pulled in its wings and, falconlike, stooped toward the ground. For a moment I thought it had detected another vole, but instead the owl made a direct hit on the back of the head of an innocent rough-legged hawk perched below. Feathers flew, and the hawk never took flight, as the owl made its way farther down the dike.

I offer the possibility that the owl was experiencing something not unlike an emotion we might feel when thwarted from what we felt was certain success; the short-eared owl was experiencing something akin to anger. Not only that, but considering what it did in response to the thievery of its prey, it was relieving or taking out its emotion on an innocent bystander. Rather than kicking the dog, the owl was striking the hawk. A form of avian displacement activity? Perhaps.

2.14 Short-eared owl striking a rough-legged hawk.
The short-eared owl does not tolerate competition
on its winter hunting territory.

When an owl catches its quarry, I've wondered if there isn't some other kind of satisfaction involved beyond merely appeasing its appetite. Is it just a series of instinctual cues—feel hunger, seek prey, find and catch, eat and satisfy appetite? Like the human hunter, could there be some implicit feeling of accomplishment in its success? As a counterpoint to a bird's feeling of anger or frustration, can they also have a sense of satisfaction, even well-being? Certainly, as in our experience, when an owl slakes its thirst and satisfies the appetite, the brain will communicate a feeling of relief, if not satisfaction. It's logical to suppose that successfully carrying through on its instincts would result in a sense of accomplishment. My barred owl, Buttons, though unable to fly, would nevertheless pounce on brown rats that came innocently into his cage to scavenge. The owl wasn't hungry, and after making the capture he would sometimes take the open perch in his aviary and hold the dead rat in his talons, as if showing off a trophy. He would hold this pose for a half hour or more before eating a portion of the rat and then caching the rest.

In late winter, the woods around my home have been flooded with the soft hooting and trilling of male western screech owls, high on testosterone as they compete for the attentions of the female birds. I can't help but think of a classic Italian opera, where the lovestruck male suitor sings his heart out to the damsel he hopes will appear on the balcony.

Along with the singing, the emotional male employs an array of other strategies to win the mate and secure a pair bond. The two birds preen each other over extended periods, and the male brings gifts to the female, typically such delicacies as a mouse, a songbird, or a crayfish. This gesture can entice her consideration of a nearby nest site, while also demonstrating his fitness as a supplier of sustenance for her and the family they might raise together.

It's reasonable to imagine that throughout the courtship there is some physical pleasure experienced at different intervals, stimulating the owls to continue. The culminating moments of mating certainly release, as they do in human coition, endorphins that are associated with a feeling of well-being.

Most of the owls I've spent time with seem to enjoy bathing, and again, it's easy to suggest parallels with our pleasures in a cleansing bath. It's not always a convenient arrangement for most owls to bathe, and water soaks their soft plumage, making flying difficult should they be confronted by a threat. Nevertheless, most of the owls living in my company—saw-whet, pygmy, western screech, barred, and barn—were quick to take to a shallow bath on warm and sunny days. It's not my impression

2.15 Western screech owl male with a gift of a mouse.
Male owls will routinely bring prey to the female during courtship.

that the birds knew all the right bathing moves instinctively, but they seemed to get in the mood as I sprinkled the water with the garden hose into their bathing pan and over them. Stepping into the bath, the birds would first cautiously dip their bills into the water, then crouch down, while shaking vigorously, to put one side then the other into the bath. On warm days Buttons would stand for a quarter hour in the bath, close his eyes, and doze. The parallels with the delights in the human bathing experience seem inescapable.

Without running a brain scan on the owl to determine what centers are most active during these moments, it's difficult to say with certainty what level of physical satisfaction the owl is getting from the bathing and the following drying out and preening that follows.

Owls are not simply creatures of instinct. They surely learn what to fear as their lives develop. Young birds that have grown up under my care responded with indifference or interest as I entered their enclosure. A wild bird that I have sought to rehabilitate will employ a range of fear-induced responses if unable to camouflage

2.16 Western screech owl fledgling confronting a Cooper's hawk. Owls have an innate threat display when confronted by danger, but will, if circumstances permit, also seek to camouflage themselves with their postures and feather positioning.

itself. Wild owls have learned to see humans as a threat, and when a lumbering form towers over them, they first scamper for an escape route. If cornered, they bend over facing the ground, with wings forward and spread to their sides, to look as large as possible in an effort to keep the threat at bay. With their nictitating membranes closed over their vulnerable eyes and their bills snapping, they are ready to fight for their lives. To this point, the bird has sought to avoid me. If contact is inevitable, it resorts to thrusting out its taloned toes to take the adversary head on.

It may be a bit of a stretch to suggest that a distraction display enacted by a long-eared owl I studied in Washington state's Potholes region was a calculated strategy,

2.17 Long-eared owl in distraction display.
When confronting a threat at their nest is ineffective, some owls will assume postures of distress or injury in an effort to distract the intruder.

rather than an instinctual response to the fear of a predator near its nest. Nevertheless, these owls do seem to make some decision whether to use one strategy or another when it comes to nest defense. Magpies are chased away, as are small hawks and other intruding owls. The bigger the animal, the more likely the owl is to try to lead the animal away. It was interesting to note, however, that the owls never bothered to either attack or distract the big grazing cattle that wandered into the shade of their nest tree, but seemed to classify them as a non-threat and paid them little attention.

I once had a burrowing owl that imitated the sound of a rattlesnake, which may have been an instinctual capacity. The owl produced the sound only when it had retreated to its burrow and I couldn't see it. Not seeing the owl made the sound far more effective in keeping me from putting my hand into its retreat. This rattlesnake call could have some effect in discouraging an invasive ground squirrel or a digging badger intent on predating the owl's nest.

2.18 Burrowing owl at badger's digs. Burrowing owls will inspect the inactive burrows of badgers and ground squirrels as possible locations to fashion their subterranean nest sites. This location will not be suitable.

The nature writer John Burroughs described an eastern screech owl that, by feigning death, sought to avoid being removed from its roost in a hollow apple tree. Although this strategy didn't work with Burroughs, we do know that some species when attacked can appear to be dead, which is a means of escaping predators, because they can lose interest in subdued prey. Barn owls, too, have been reported to feign death, and the moment the perceived predator is distracted or becomes uninterested, the prey comes to life and makes its escape.

Some of the owl nests I have examined have had their edges laden with surplus prey. A brooding female long-eared owl on a flattened magpie nest was surrounded by a circle of stacked carcasses of wood rats, voles, and the odd lizard. Her young were just hatching, and there was no immediate need for this largess. It's not a case of greed, but rather the behavior of the prey-supplying male to take every opportunity to capture something for the family. The owl cannot afford to make measured

2.19 Snowy owl perched at Puget Sound shoreline.

harvests, although many owls will cache prey in locations separate from the nest when the need for it is not immediate. The habit of caching prey about the nest is not a trait that is likely to be passed on. Other predators can be attracted to the resulting smell of accumulated uneaten meals.

I have seen young western screech owls appearing to play with clumps of moss they grab with their toes. Picking at it with their beaks, they reach up to grip it with a foot, then drop it over the edge of a branch to watch it fall to the forest floor. There's no attempt to consume it, only to reach out and pick it up and, after a moment of interest, let it fall. Perhaps this innocent behavior is practice in grabbing an object, preparing the owlet to capture a meal, and indeed this rudimentary play does resemble a threat-free preparation for the realities of life.

More than a million viewers on YouTube have watched a more obvious engagement in play by an owl. This video, made in Tarragona, Spain, in 2011, shows a barn owl and a cat playing, with all indications that both species are engaged in a mutually rewarding frolic. The owl, named Gebra, flies up and down a rural path with the cat, Fum, in pursuit. The owl swoops low over the cat, and the feline leaps up beneath the bird just short of a collision. When the owl lands on the ground in the path of the cat, the mammal races toward the bird as if to pounce, only to jump completely over it. If the cat is sitting indifferently, the owl is quick to fly over to nuzzle the cat, treating it like it might another owl it was seeking to preen affectionately. After a breath-catching timeout, they are back at it again, with Fum clearly keeping its playful moves within the range that Gebra can tolerate.

To survive, every higher-order species takes an interest in its environment in a manner that transcends a knee-jerk response to an obvious stimulus. Sights and sounds are filed in the memory bank of the animal, with some requiring more attention than others. When watching the wintering snowy owls, at first they appear as an indolent stack of white feathers, but a longer look through the spotting scope will reveal their alertness. Although their eyelids are closed to mere slits, you will nevertheless see subtle turns of their heads as they adjust their line of sight to overflying birds, or position themselves to better hear a nearby sound. Where a moment before the bird appeared to be asleep, if it hears a sound, its eyes snap open and fix on a particular spot in the grasses nearby. If a noise suggests a vole prowling about on the other side of a drift log, the owl may lift off to perch above the sound to watch and wait. If its curiosity leads to reward, the big owl drops down and makes a capture.

From perches above a creek, western screech owls watch the darting underwater shadows and then, should the action look promising, move in closer for a better look. Often it is just a ripple of light, but it sometimes leads to the discovery of a crayfish that the owl can capture in the shallows.

Just as familiar sounds can trigger mental pictures and associated memories in our brains, the owl can remember sounds that have become important to it. I have watched our resident male western screech owl dozing midday near its nest when a Swainson's thrush gave forth with its melodious call from shadows in our woods. Up to this point, the owl had appeared to be sleeping and was indifferent to the chattering of chickadees and nuthatches. The thrush's song, however, seemed to pull a trigger on the owl's memory, that this was a prey species worth pursuing; its eyes now open wide, the bird turned and awaited another cue. Before the thrush had finished its song, the owl was flying off into the woods in its direction. Within minutes, the gray-brown male owl made a loop out into the sunlight before flying up into the nesting cavity to deliver the captured thrush to its awaiting mate and nestlings.

Defending a territory from competing pairs of owls is often part of the ritual of establishing a nest site and raising a brood. A good cavity for the western screech

2.20 Western screech owl discovers crayfish.
Owls learn to investigate the slightest indications that prey might be present.

owl, boreal, saw-whet, flammulated, or pygmy owl is worth fighting for. When the hormone-driven bird is in full gear, a lot of energy is put into securing and protecting the breeding and hunting grounds. There are many stories of curious naturalists who have climbed up to great horned owl nests, unprepared for the fury the adult owls unleash in protecting their young. The time one of "my" western screech owls went after a member of a marauding mob of crows—all of them much larger than the owls—was the most dramatic display I've seen of owls defending their nestlings.

The writer and epidemiologist Marston Bates recounts an occasion when he grew frustrated with audience members asking what value this or that species had in serving the interests of humankind. When someone questioned the purpose of a particular insect, it seems Bates lost his patience and asked the inquisitor, "What are you for?" Although his response may have been abrupt, I have a sense of how the scientist was feeling. I, too, get impatient with an attitude that is only interested in justifying the

2.21 Western screech owl striking a crow. Adrenaline overrides any reluctance for a parent owl to attack when crows threaten the fledglings at a western screech owl's nest.

existence of another life form on the basis of some obvious benefit it might provide us. The most important insect in Bates's work, which had a tremendous positive effect on human health, was the mosquito. His study of the insect's life cycle led to great advances in the epidemiology of yellow fever in the 1930s and '40s. Most scientists would tell us that we are only now beginning to understand the complex direct and indirect interactions between species that are essential for the vitality of ecosystems.

As we know, in spite of evidence to the contrary owls, hawks, and eagles were considered vermin well into the twentieth century, and were routinely killed. Predatory birds by the thousands were giving up their stomachs to science and proving that the prey they consumed were the very rodents and insects that were the bane of our agricultural interests. It's perplexing that even to this day pockets of ignorance remain. Domestic and feral cats, along with a long list of nocturnal mammals, are far more responsible for declines in populations of songbirds and game birds, but it is still the bird of prey that gets much of the blame.

One revealing study in the early 1900s of barn owls nesting in a building associated with the Smithsonian Institution in Washington, D.C., revealed that, during the breeding period of one year, the family had captured and consumed at least 450 small rodents. (It is possible to get this information by analyzing the birds' castings.) Although three-quarters of the diet was composed of mice, almost all of the rest was rats. One can assume that most of these mammals were former occupants of government grounds, but, despite the clear benefits in pest reduction, little action was taken to pass laws protecting the birds until far later in the century.

It was nearly another hundred years before an enlightened understanding of barn owl services dawned. The Barn Owl/Rodent Project of 2011 in California is of particular interest. It not only demonstrates the importance of these birds in controlling agricultural pests, but also shows how we can work with the owls to apply their rodent-hunting skills to our purposes.

Nesting boxes for barn owls were placed in a hundred-acre vineyard that was heavily infested and damaged by pocket gophers and voles. Within a year, more than a hundred owls, including breeding pairs and their young, were established on the site. The impact was immediate and significant: estimates were that in two months of relentless hunting, the owls had caught and consumed some seventeen thousand rodents. As a result, crop damage in the vineyards diminished dramatically.

Even though a fledged barn owl has but a one percent chance of reaching a potential age of ten years, they do breed by their second year, and a pair of owls with five owlets will consume some three thousand rodents in a nesting season.

The importance of owls in rat control in the suburban neighborhood where I live emerged when barred owls began to occupy and breed in our woods. The resident screech owls were not consumers of rodents as large as adult brown rats (*Rattus norvegicus*), and neither the barn nor great horned owls found our riparian location suitable for residency. For many years after we first moved into our older home, rats were a common and persistent problem. House rats (*Rattus rattus*) even invaded the duct system for our oil heat, and the brown rats chewed their way into some of our furniture and routinely scrambled out of sight when the basement lights were turned on. With a female rat producing five litters of eight pups a year, we were running to catch up as we constantly set and baited traps. Then, ten years ago, barred owls took up residence in the woods in some numbers, and the rat population began to diminish appreciably. Examination of the owl castings makes it clear why. Although the barred owl has also discouraged the presence of the smaller owl species, we now have, at no cost, a resident pest control presence.

When it comes to augmenting our soils, we are inclined to quick fixes and routinely turn to commercially produced fertilizers. Our agricultural production successes require it. Over centuries, nature applies nutrients in a far more measured and subtle fashion. I think particularly of how, when our spawned-out salmon in the Pacific Northwest die, their remains are dispersed throughout creeks and larger waterways, and some of their carcasses are transported into the forests by scavenging bears and eagles. The nutritious remnants of fish flesh, bone, and cartilage they leave behind end up nurturing aquatic animals and future generations of fish, and become the building blocks of the Pacific Northwest forest. After its three years of life in the North Pacific, the salmon still delivers its nourishing body back to the streams and forests that made its life possible. It is a classic example of reciprocity in nature.

Owls, too, should be considered dispersers of nutrients, particularly nitrogen, because their excrement and castings are spread through their habitats. Over their lifetimes and those of their offspring, this application and circulation of nutrients for plants can be considerable. In turn, we are the beneficiaries of this interdependency of the plants and animals of the forest. Healthy stands of trees provide water filtration, hold soils intact, inhibiting erosion, and retain water that might otherwise flood areas or contribute to the triggering of catastrophic landslides, such as those experienced in western Washington in 2014. The carbon capture, air filtration, and noise abatement that the forest provides is profoundly important to our living in a healthy environment.

In their given haunts, owls occupy unique positions. To scientists, they are a special indicator species: their presence or absence suggests the condition of the habitat. An old-growth forest in the Pacific Northwest, for example, is complete if there are spotted owls present. They are uniquely suited to thrive in these diminishing and unparalleled locations. When the birds are able to maintain their occupancy, it is a significant indication that there is a proper balance of old-growth species and conditions of biological diversity sufficient to support the birds' feeding, breeding, and roosting requirements. Such conditions for any remaining ancient forests throughout the world are becoming increasingly rare, and, not surprisingly, the owls that are adapted to live there are in turn disappearing. As these forests diminish and vanish, so do all of the yet-to-be discovered complexities of species relationships and biological diversity within them. For the sake of science and ourselves, it benefits us to sustain them. Otherwise their worth, both obvious and subtle, will never be determined.

2.22 Spotted owl in an old-growth forest. The spotted owl is one of several ancient-forest owls around the world that are threatened by the destruction of their habitat.

2.23 Western screech owlet in a person's hand.
The future of many owl species will be determined by our stewardship.

Given their nocturnal habits and retiring manner, owls are among the least well known birds of the avian kingdom. They represent to science an endless array of mysteries to be solved. In the process of doing so, we will undoubtedly find yet another reason to cherish and practice responsible stewardship. Judging by their exceptional abilities to see, hear, feel, and know their worlds, so different from ours, we have much to learn from them. And, as my friend the artist Fen Lansdowne used to say in response to the question of why he was so interested in owls: "Because they can fly, sir, and I cannot."

Implicit in the belief that we are rational beings is a duty to respect other rational beings, both present and future. Without debating the capacity of an owl to exercise rationality, we can argue that the species has provided efficient ecological services that are essential to our welfare. Although I personally don't believe our generations have the right to be indifferent to the welfare of these birds, I do think every ethical person would agree we have no right to deny future generations of our species the presence of owls in their lives. Beyond the obvious control of pests that damage our agricultural and even our domestic well-being, owls have among the oldest and most extraordinary legacies of inspiring the creative mind in literature and the arts. When studied in the sciences, there is no end in sight. They will continue to inform us in matters of how the world works. Knowing and appreciating these species is one of the portals for us to enter and understand our place in nature. For our own sake, we

2.24 Western screech owl fledglings observed by children.
Growing up with families of owls for company provides real-life learning that vicarious adventures through the media can never provide.

have an ethical responsibility to protect and sustain their lives, thereby providing future generations the thrill and benefits of knowing and living with them.

It is ironic that there is so much talk and solicitation for funds to clone present species by a prolonged and expensive process, with methods that may eventually restore extinct species to life. It seems to be a divergence, when a convergence is needed. Cloning can concentrate our attention in the laboratory and suggest we can remedy our blunders in degrading the environment and reducing species diversity by bringing back to life an extinct animal. What's needed is a focused effort of science and public support to salvage and exercise stewardship over what little remains of our natural history. A rabbit may be pulled out of a magician's hat; but, more important, where will it live and what will it do? The world that the passenger pigeon and the ivory-billed woodpecker inhabited is long gone, but we still do have places for owls.

THREE

Owls and Human Culture

There is a scene from the 1997 film *As Good as It Gets* where an artist (played by the actor Greg Kinnear), having experienced a series of serious physical and emotional setbacks, is suddenly revived to a creative fervor by the lines and shapes of an undeniably beautiful form. His aches and pains are forgotten in his rush of determination to hold the moment of what he's witnessing. "I've got to draw you," he says. "You're why cave men chiseled on walls." The line rings true. Words can't do justice to our passionate reactions to what we see or feel, so we fashion an artistic response to pay homage to and clarify the moment.

Judging from the archaeological evidence, we can safely say that owls are another reason why cave men chiseled on walls. In a collection of stylized renderings of ice-age mammals made thirty thousand years ago, there appears a European eagle owl (*Bubo bubo*). Placed near the roof of a cave at Chauvet, France, it faces the observer with feathered tufts erect and eyes wide. I can imagine some of the motivations that this artist of the late Pleistocene Age might have experienced, and they were not altogether different from my own artistic inspirations. Seeing this very imposing, powerful, and silent flying creature that comfortably occupies the dangerous night touched something beyond utilitarian concerns. There was a stirring of his nascent aesthetic sense. In his emotional efforts to describe its forceful beauty, he fashioned a permanent depiction of it.

A sampling of later owl-inspired works and references includes an ancient Egyptian sculpture of a barn owl's face that is rendered with such great precision, one can

3.1 Eskimo wood carving of a snowy owl.
The Eskimo artist Buckland carved this owl in wood with inlaid old ivory, intending the work to be hung in a dwelling.

easily see the interest the sculptor had taken in this subject. Although not named as one of the Egyptian gods, these birds were mummified and pictured hieroglyphically.

More than two thousand years ago the owl was employed, for educational purposes, in a very sophisticated way in Sanskrit beast fables known as the Panchatantra tales. They described ongoing conflicts between Cloud Hue, King of the Crows, and his owl enemy, Foe Crusher. This informative allegory is based on a sound understanding of the mutual animosity existing in nature between the crow and the owl. The narrator of the fable suggests strategies for the Crow King when dealing with his enemy the owl. These included an understanding of the arts of conciliation, confrontation, withdrawal, diplomacy, and deception.

The owl's status among the ancient Hebrews was far less elevated. In Leviticus 11:13–17, the owl and other predatory birds are named as abominations that are not to be eaten. Seen in association with their prey and often occupying ruins, they were deemed unclean. A biblical reference to owls in Isaiah 34:13–15 severely condemns an enemy to ruination by sentencing them to a "habitation of dragons and a court of owls."

The Greeks, on the other hand, saw an expression of erudition in the owl's mysterious and remote manner. Their goddess of wisdom, Athena, is depicted with an owl on her shoulder, or wearing a helmet with an owl symbol emblazoned on its side. An Athenian coin, the silver tetradrachm of the fifth century B.C., also had the image of the little owl (*Athene noctua*) struck on one side of its face. In fact, there are numerous Greek coins of various denominations that feature the image of an owl. Still, the importance of this owl metaphor went beyond the marketplace. When the Greek army was assembled for battle, it was believed that a victory could be anticipated should an owl fly over the troops. Rumor had it that a caged owl was kept ready for a quick release at these times should a wild owl fail to show up.

Borrowing yet again from the Greeks, the Romans adopted the goddess Athena, renamed her Minerva, and soon altered the meaning of the owl as well. Early Romans associated the vocalizing of the owl with predictions of death. History has it that the demise of emperors and such Roman luminaries as Julius Caesar, Augustus, Commodus Aurelius, and Agrippa were all preceded by the calling or the presence of owls nearby. In the Roman agricultural community of the time, the superstitious country people hung up the bodies of owls to avert storms.

A cursory look into myths that exist around the world reveals that more than seventy different countries, from Abyssinia to Wales, hold folklore traditions related to owls. A few of these include tribal cultures in Africa that believe if a large owl is about a house it is an indication a powerful shaman is inside, and the owl is a messenger between the shaman and the spirit world. There is an Australian aboriginal belief that owls are a form the human soul can unite with at death. Ancient Arabs saw owls as the spirits of individuals who had been unavenged when murdered and the cries of the birds were pleas for vengeance.

Judging from the scale and detail of the images of owls etched into the cliff at Nine Mile Canyon and other locations in Utah, one would have to conclude that the bird had made a significant impression on the pre-Columbian artists of the Fremont culture. The myths of the Hopi of the American Southwest regarded the burrowing owl as a sacred bird. Because of the bird's fondness for underground burrows, the Hopi people judged it to be a symbol of the God of the Dead. Such a subterranean life gave it direct contact with the underworld and the activities of the soil. The Zuni people considered the owl feather to possess talismanic powers. Placing it in company with a baby would keep the evil spirits at bay. To this day, the First Nation People of the Pacific Northwest carve the owl figure as part of their totemic crests to pay homage to ancestors.

In the far north, Eskimo artists carry on ancient traditions with their interpretive carvings, particularly of snowy owls, in ivory, whalebone, wood, and stone. These determined and inventive artists worked with the materials at hand. The individual vertebrae of the bowhead whale became the head and body of the snowy owl. The massive transverse processes extending from the vertebrae were shaped convincingly into the open wings of the bird.

In the western world throughout the Middle Ages, the owl was looked upon with a mixture of fear and admiration. In England, owl bodies were nailed on doors of a house and on the barn to keep evil spirits at bay and to protect the livestock inside. Here, however, the Greek take on the owl's sagacious nature had its influence and was incorporated into the Arthurian legends. The wise wizard Merlin is described as having an owl upon his shoulder, and throughout this period the bird is seen as a companion of seers and alchemists. In France, it was believed that when a pregnant woman heard the call of an owl, her child would be a girl. (I have particular interest in this myth. As the father of four daughters, my wife and I heard owls calling continuously while our children were conceived, born, and raised.)

Along with the early integration of the owl's metaphysical aspects into the cultural myths and literature of the people came a curious application of the bird's physical parts. Various recipes were proposed as medicinal complements to the welfare of

3.2 Inuit man-owl of stone. Referencing a snowy owl, the carver Latcholassie from Cape Dorset has chosen to explore the humanlike aspects of his subject in this stone sculpture.

humankind. Their eggs received special attention. In the Cyranides, a fourth-century compilation of magicomedical works written in Greek, it was claimed that a soup made of owl eggs under a waning moon was a cure for epilepsy. The disputable wisdom of the character Apollonius in Philostratus' *Life of Apollonius* includes the recommendation that eating an owl's egg will help one develop a dislike for wine. In parts of Europe this suggestion was taken even further: eggs were not only considered a proper cure for drunkenness, but also, should a child be given an owl's egg, the youngster would never become an inebriant. In 1635, John Swan, writing in his *Speculum Mundi,* included a list of all sorts of maladies that could be cured by owl parts. A few recommended using its feathers for a case of gout, the blood to treat paralysis and hair lice, bile for bedwetting, and bone marrow for a migraine headache.

3.3 Barn owl nailed to a barn door. In Europe,
well into the nineteenth century owls were killed and then nailed
to the doors of barns and homes to ward off evil.

3.4 Coat of arms for the borough of Oldham, England, featuring owls. The Latin phrase exhorts one to have courage, think for oneself, and commit to tasks that need to be done.

Owls are featured on the heraldic signs of cities, boroughs, and individual families. Particularly in European countries, the owl image is emblazoned on a shield format to convey the power, intelligence, and fair nature of its owner. Among the coat of arms for the cities of England, that of Leeds stands out. Rather than the often-seen fantasized forms of unicorns or lions, we see three stalwart owls looking straight ahead to confront the observer. In 1635, knowing of the mythic standing that owls had among his countrymen, George Wither of Oxford College produced his book *A Collection of Emblemes, Ancient and Moderne.* He portrayed the owl as symbolizing, among other aspects, various virtues or conditions. An owl was pictured perched on a book to represent wisdom, looking content amid the chaos of being mobbed by crows to convey calm, and roosting atop a human skull to symbolize mortality.

Leave it to Shakespeare to seize on the owl and lavishly use its standing in myth and folklore as a dramatic metaphor to advance the action of his plays. A few of many such examples include Act 1, scene 3, of *Julius Caesar.* The playwright sets up the impending death of Caesar by having Casca proclaim, "And yesterday the bird of night did sit even at noonday upon the market place hooting and shrieking . . . I believe they are portentous things unto the climate they point upon." In the fourth act of *Macbeth,* the owl appears as the harbinger of evil and doom. One of the witches, Harpier, is transformed into an owl, and the shrieks of the bird are heard offstage when Duncan is murdered. To further suggest Macbeth's depravity and acceptance of his fate, the Bard has his character so inured to the owl's prophetic calls that he exclaims, "The time has been my senses would have cooled to hear a night shriek . . ." And, as with so many colorful phrases that we still use routinely today, Shakespeare was among the first to effectively employ the term "night owl." To frame the tension in *The Rape of Lucrece,* he uses the line "The dove sleeps fast that this night owl will catch." And in *Richard II,* "For night owls shriek, where mounting larks should sing."

With the advent of the Renaissance, sculptors and painters began to incorporate the owl's image to leverage its metaphoric value. The skills that were applied to the bird's interpretation gave the viewer a deeper appreciation of the species' beauty and dignity. To some degree these interpretations began to move the birds out of the realm of mystery and into the light of reality.

Michelangelo found an opportunity to carve a stout and convincing barn owl in marble. The bird is contained in a larger sculpture titled *Night,* carved for the tomb of Giuliano de Medici. Tucked up into the shallow recess of a reclining female's crooked leg, the barn owl occupies the outer edge of a darkened space that is something of a vestibule, similar to what an owl might occupy in the day, then appear from at night. Although the carver has given great care to define the species by its form in general, he has paid particular attention to the shape and patterns of the unique feathers of the owl's face and the manner in which the wing plumage is layered. With one foot extended as if to emerge from the darkness, the owl has its other leg perched higher up to give support for its movement. The artist may very well have had a barn owl in his studio for reference, and he clearly was familiar with the habits and haunts of his subject as well.

An examination of Hieronymus Bosch's paintings displays the form of owls bearing witness to the chaotic events around them. Their position suggests their role in weighing judgment on the action, and giving art scholars many interpretive possibilities to chew over. As a young artist Albrecht Dürer sought to interpret the owl more as a subject of beauty, rather than to exploit its symbolic potential. By investing his talent and interest in his subject, he created a portrait of the little owl that to this day stands with the finest direct artistic interpretations of this subject ever done.

The owl has been the subject of children's rhymes, perhaps in response to what appears to be its rather pompous and almost arrogant appearance. Among the most well known is Edward Lear's stanza from his *Book of Nonsense:*

> The Owl and the Pussy-cat went to sea
> In a beautiful pea green boat,
> They took some honey, and plenty of money,
> Wrapped up in a five pound note.

In *Punch,* the British weekly magazine of humor and satire, an anonymous short poem is both satiric and wise:

> There was an old owl he lived in an oak
> The more he knew the less he spoke
> The less he spoke the more he heard
> O, if men were all like that wise bird.

Approaching the matter of how and where the owl emerged from being mythical to being real, it is important to note how the Japanese artists of the Edo Period depicted these birds. Over the span of more than two hundred and fifty years (1615–1868) the artists produced work influenced by an amalgam of Taoist, Shinto, and Buddhist beliefs along with a hint of animism. They integrated the owl's form into compositions of the bird's habitat. Be it woodland, streamside, or in a vast canyon, the subject was painted as part of the entire natural community. One senses in these extraordinary depictions that the artists were seeking to be like an owl and come into an accord with nature. The bird and its surroundings were one.

By the early nineteenth century the owl's reputation in Western thinking had taken a more objective turn, particularly as the arts had joined science in exploring all things natural. The Darwinian age had been set in motion, and rather than being cloaked in myth and fear, owls were something to be studied and celebrated. The naturalist artist Alexander Wilson, and then John James Audubon and John Gould, advanced the knowledge and appreciation of the birds of the planet. The animated

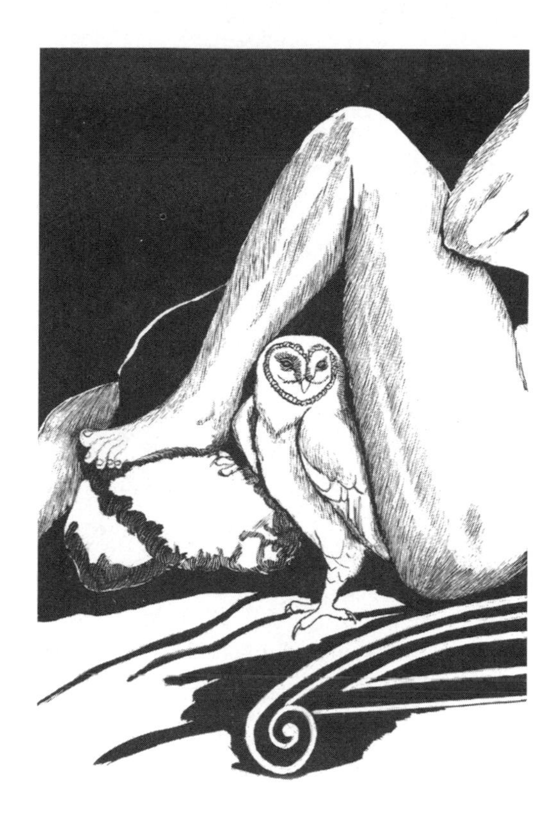

3.5 The barn owl from Michelangelo's *Night,* carved between 1526 and 1531. The superbly rendered owl metaphorically supports the sculpture's title.

and superbly designed masterpieces of Audubon, who was influenced by the Edo Period artists, conveyed as never before the range of scale and beauty possessed by the North American owls.

In the United States, among the first of the writers to describe these subjects with poetic eloquence was Henry David Thoreau. He wrote in his journal of the great horned owl's call sounding like a primal church bell as it rang "far and wide, occupying the spaces rightfully . . . it is a grand, primeval, aboriginal sound." A half century later, John Burroughs, a popular nature writer of the time, described the owls along his farmstead. His respectful and inquisitive narrative on owls encouraged readers to discover the nature of these birds for themselves.

Science was on a quest throughout the world, observing and bringing back specimens to expand the understanding of the company we keep on the planet. By the beginning of the 1900s the diversity of the order Strigiformes expanded dramatically, as these reclusive birds were collected and described from the widest range of locations and habitats for study.

Even before the middle of the twentieth century, owls were finding a presence in the film arts of the Western world. Disney Studios incorporated the wise old owl to give gravity to their cartoon stories and later produced natural history films that spotlighted the lives of some of the species. Horror films for both adults and children inserted the images and sounds of owls to create an atmosphere of suspense and mystery. What was sometimes distracting to anyone who knew owls well, however, was that the sound technicians rarely got it right. In vampire films in particular, the recorded calls of owls, used to set the atmosphere of mystery and terror, were often the cooing of ring-necked doves. Most likely only a few people in the audience noticed, and these ersatz owl calls were still effective in creating a frightening atmosphere.

When I watched the 1973 supernatural horror film *The Exorcist* I was astounded to see in it an owl image taken from a book by my friend the bird portrait painter Fen Lansdowne. The director, intending to add to the chilling atmosphere of the moment, allowed the camera to focus in on the looming body and full face of the great gray owl from Landsdowne's *Birds of the Northern Forest,* volume 1. Alone, Lansdowne's rendering of the owl is a stunning interpretation, but when placed in the context of the film's story it took on a more sinister quality. I don't believe Fen ever saw the film, and he wasn't much perturbed about its use.

J. K. Rowling's seven Harry Potter books and the subsequent films used owls to great effect as a source of knowledge. Seeming to take a cue from the wizard Merlin of Arthurian legend, Rowling invests all manner of owl species to provide the communication network among her characters. Hedwig, a snowy owl, is Harry Potter's trusted associate who keeps him in touch with other key characters in the stories. Great gray, western screech, scops, and an eagle owl are all associated with characters from the books and films. In a nod to the historical idea that owls are fonts of knowledge, a team of them is used to deliver the *Daily Prophet* and *Quibbler* newspapers.

Today owl images are beautifully presented in paintings, sculpture, and photographs on an unprecedented scale. To list and describe the contemporary artists, living and dead, who have competently and originally interpreted the owl is an undertaking so vast that it would require another book. Suffice it to say that from

3.6 Plaster rendering of a great horned owl.
Fashioned directly from an owl specimen, a plaster bust of the bird
serves as a model for a life-sized bronze casting.

realism to abstract expressionism, or from the vast impressionistic canvases of the Swede Bruno Liljefors to the abstract works of Picasso, Miró, and Kandinsky, owls have been interpreted artistically to convey messages and inspiration for the ages.

I'm certain that owls will sustain their presence in our cultural evolution, and in particular continue to influence our artistic expressions. Their influence on me, as a writer and artist, has been a long one and provided the momentum and approach that made this book possible. With this in mind, I was curious to determine if I could describe in some tangible detail just what it is about owls that has inspired and motivated me to devote much of my artistic career to them.

It wasn't long after I moved to the Pacific Northwest before I went out into the woods to discover what lurked there. The owls were waiting. I was in my early twenties and this was the beginning of the decade of the 1960s. Along with the many issues confronting an idealistic college student were those early concerns over a polluted and degraded planet. Rachel Carson's book *Silent Spring* spoke to the threats that agricultural chemicals were having on wildlife, and birds of prey were particularly vulnerable, as they accumulated the toxins. Owls were part of this group at the top of the food chain, and I imagined that if I didn't study and draw them immediately, the possibilities of extinction were such that I might never have another opportunity.

Fortunately, with the enactment and enforcement of laws and public information, birds of prey weathered these threats for the most part. Although diminished in numbers, they have sustained populations. What I did discover, however, was that my initial attempts to replicate the birds' appearance opened up the more subtle and intriguing aspects of them. I also believe that these early works, when shared through publications and art shows, transcended words that often led to intense exchanges over the fate of the birds. At the time I was only a modest participant in what other artists across the country were doing in giving the public a picture of an attractive and engaging species that deserved our interest and attention. I have come to believe that this emphasis on aesthetics rather than debate helped to foster a more informed electorate. It contributed to a climate where emotions settled down and a reasoned discussion ensued.

The perilous ecological position of some owls is only part of my attraction to them. Owls remain mysterious, and I draw and sculpt them not so much because of what we know about them, but because of how much more there is to discover of their nature and behavior. As discussed throughout this book, their habits keep them at a distance where they remain elusive. Their colors are subdued, and the patterns in their plumage are such that they match their surroundings and are not easily seen. Compared with more accessible species, we know little about their behavior. All of this becomes a magnetic invitation to engage with the birds and learn more. A creative response sweeps one away into fresh discoveries and greater appreciation. The adventure is so intense that the sharing of the results is sometimes only a secondary pleasure.

The physicality of owls begs for a creative response. Putting pencil to paper or chisel to stone is my quest to get closer to reproducing what it is to be an owl. I set out to reveal this "owlness" that sets these birds apart from other avian species. The forthright manner by which owls can confront us, with their large eyes set into broad faces with heads atop upright torsos, makes them appear among the most humanlike of birds. When they look back at us from the darkest recesses in nature, they remind us of our origins. This is a call to explore the details of the birds' unique countenance. I'm fascinated by the facial features, including matching tracks of ridged feathers curving inward from either side of their beak. There are the broad shields of specialized feathers that encircle their eyes and simultaneously gather all available light and surrounding sounds. Their toes, whether thickly feathered with long lacy plumage like the snowy owl or naked and elongated like the barn owls, are elegant, sharp-taloned designs that speak eloquently to their function.

Along with the owls' singular shapes and contours is their plumage, so subtle in hue and pattern. Designed for a secretive life, these birds lack the stunning intensity of color and contrasting plumage patterns of orioles and cardinals and blue jays. This muted appearance possesses attractive gradations of warm color and complex patterns that are a compelling challenge for an artist to portray.

Getting close to the owls in my life provided me with yet another range of appreciation. Like any sentient animal, their postures and gestures reflect emotional states. I have sought to develop interpretations of owls that are more than an illustration to convey the subject's appearance and scale. Knowing something of their behavior, I'm motivated to shape my subject to a degree that does justice to their emotional state. I believe owls are inquisitive, playful, wrathful, determined, and even contemplative. There is an artistic imperative to try to convey these owl behaviors in my art with an authenticity. Avoiding anthropomorphizing the subject is a challenge, but the more we learn about our avian companions the more we discover what we have in common with them. As long as we have the pleasure of keeping company with owls, they will continue to stir our imaginations, inform us and invigorate the artists among us to celebrate them.

Angell

FOUR

Owls in Company with People

The common way to group owls is by genus, which does provide a clearer picture of relatedness among them, but there are other ways of considering these diverse species. They can be categorized by size, and indeed some are very large while others are medium and small in their proportions. To some degree they can be grouped by how they hunt and the food choices they make. For my part, I am intrigued by considering these species according to where they can be found and in particular how some owls, coevolving with humans, have managed to live in rather close association with people.

Throughout much of the world some species of owls have managed to share their habitat with humans. Over thousands of years they have not only developed a tolerance for our activities but also learned to exploit the feeding and breeding habitat that we, without special intent, provide for them.

Perhaps one of the most successful owls that chooses to keep us company is the barn owl. Across North America, Europe, and Asia these birds occupy spaces where our structures furnish nesting sites and our habits assure them of a regular supply of small mammals. Both the eastern and western screech owls live in our suburban settings and find ample food sources and, should woodpecker cavities be absent, take readily to nesting boxes to raise their families. Northern saw-whet owls live near us for the same reasons in our wooded neighborhoods, and both the short-eared and long-eared owls, while also occupying wilder regions, comfortably hunt the open fields and grasslands adjacent to our settlements. The aggressive barred owl has

4.1 A pair of barn owls perched in rafters.

expanded its range across North America to take up residence in our local parks, and the ubiquitous great horned owl, while something of a symbol of the deep and impenetrable woods, has moved into suitable habitat within cities.

Secretive and well concealed, these owls occupy spaces near us, and few of us notice. They render immeasurable service in suppressing pest populations, and when we hear or see them, they inspire. They do have their limits, however, and sustaining their diversity and numbers requires thoughtful stewardship of habitat and cautious use of any toxins that could threaten their health.

Barn Owl (*Tyto alba*)

By my early teens, my family had moved from the floor of the San Fernando Valley into one of the canyons of the adjacent foothills. Here, the barn owl's raspy, high-pitched scream, along with the *"ahoo—hoo-hoo"* of the great horned owl were a routine part of our summer nights. The barn owl's harsh proclamations, whether passing overhead or perched in some dark recess of a palm tree, building, or stand of eucalyptus trees, really fired up one's imagination. By middle school my chums and I were setting off on bike hikes to discover the nests of the owls, thinking it would be an adventure to raise one of the owlets. It was.

Our quest was given some focus when one of my pal's fathers, who worked in the Hollywood film production field, mentioned that owls were nesting in the motion picture sets at some of the local studios. This possibility was enough to set us off on a series of trespassing expeditions over the back fences of Universal and Republic Studios where old film sets, unoccupied and rarely used, seemed like suitable locations for nesting barn owls.

It was in one of the Old West town structures, with its false fronts for the sheriff's office, general store, and a swinging-doored saloon, that we discovered an owl nursery. Entering the back of the saloon, we could see a barn owl roosting in the rafters over a collection of tables and chairs that we were sure we had seen in a recent western at the El Portal Theater. There was a generous spread of owl guano where John Wayne, Bob Steel, and "Hoot" Gibson had had a drink and played a hand or two. In one corner, we spotted a likely nesting site and, climbing up into the rafters, we discovered a narrow shelf that the owls had selected for brooding their family. There were four owlets that ranged in size and development—a phenomenon, as we would learn later, that is typical of all owl families, because the female begins to incubate with the first egg laid. The oldest youngster bowed, swayed, and hissed, but not enough to discourage me from picking him up, wrapping him carefully in my mother's softest bath towel and settling the bird gently into the bottom of a big paper bag.

This activity is not to be recommended. Although we were doing no damage to the private property, we were trespassing nevertheless, and the taking of a wild bird today is against the law—with good reason. Still, I look back on it as a key part of my gradual immersion into the study and depiction of these remarkable species. I'm thankful, too, for a parent who put up with the constant clamor of the growing owl, the cleanup of its enclosure, and the inevitable grief that comes with separation when, after several months, I released the owl back into the wild. A good part of my time with this owl was spent as a surrogate parent introducing the bird to live prey that it could catch so it might develop the hunting skills required to give it some chance of survival when released.

RANGE AND HABITAT

One might argue that this bird has benefited more from humankind's enterprises than any other owl species. As we have expanded our settlements and activities, the barn owl in turn has exploited the hordes of scavenging rodents that accompany us. Occupying every continent, the barn owl, coevolving with humans, has adjusted its habits to become primarily a nocturnal hunter. Although it avoids human contact as much as possible, the owl finds suitable roosting, feeding, and nesting locations amid our neighborhood structures.

A bird found worldwide in temperate climates, barn owls prefer more open habitats throughout the lower forty-eight states and southward through Mexico into Central and South America. This long-legged, rodent-catching bird seeks to take up residence in both rural and suburban settings, and if its feeding requirements are met it will occupy urban structures as well.

FOOD PREFERENCES

Barn owl castings are a mainstay of many high school biology classrooms. When properly dried, they become essentially germ free and can be dissected to reveal the remains of the meals swallowed whole by the bird. Looking at the intact skulls and fine bones and fur provides students with an appreciation of the remarkable digestive powers possessed by the owl. Moreover, the experience provides one with a stronger sense of how important these species are in controlling local pests.

Determined in part by where the birds reside and what the populations of small mammals might be, they consume a range of prey including pine, meadow, and

4.2 Range map of barn owl in North America.

jumping mice, along with assorted species of gophers and voles. The barn owl is powerful enough to take mammals as big as muskrats and spotted skunks, with an occasional bird—some as large as green herons.

VOCALIZATIONS

These owls possess a large repertoire of vocalizations. The most common include whinny and twitter calls used by the begging youngsters; a hiss associated with an alarmed or threatened owl; and the raspy scream, both a threat and a distress call often used by an adult to establish its territorial claims and maintain contact with its mate.

COURTSHIP AND NESTING

In some parts of its range, courtship for these monogamous birds begins as early as January. The female produces a clutch of five to seven eggs at one of the nesting sites the male has shown her. In a single year a pair may raise more than one brood, with the female incubating another clutch at a second nest while the male continues to feed the fledged young from the first.

Although cavities and recesses are preferred, their nest sites are rather catholic. Buildings that have been standing in a state of disuse are prime real estate, along with rocky cliffs, church steeples with bell towers, and the occasional crow's nest. Some have even dug burrows with their beaks, and they have certainly taken to the accommodations afforded by nesting boxes.

The incubation period beginning with the first egg laid continues over twenty-nine to thirty-four days. The eggs hatch asynchronously, with the first one hatching within twenty-one to twenty-four days. Within two weeks the first-hatched owlets can walk and are flapping their wings to move about the nest. At eight weeks the owlets, although being fed by their parents, are flying near their nest. In their third month, they are nearly independent and catching some of their own meals.

THREATS AND CONSERVATION

In parts of their range where great horned owls compete for space and food resources, the larger birds will take barn owls as prey and eliminate the competition. In Europe, barn owls are killed by golden eagles, goshawks, peregrine falcons, and eagle owls.

As in all owl species, if their selected habitat is compromised, their populations decline. As the outbuildings of small farms are removed in favor of larger agricultural operations and their structures, nesting and roosting sites for the owl are diminished. Without such accommodations the owl cannot take up residence and provide its rodent control services.

There is concern that a rodenticide, using the blood-thinning agent warfarin to kill rats and mice, could affect the vitality of barn owls. Owls catch rodents that have ingested the poison, and in turn the birds can accumulate the chemicals in their systems with serious consequences. Eggshell thinning, detected in some midwestern populations of barn owls, suggests that agricultural chemicals passed through the food chain may harm the birds. These conditions are not unlike the damaging effects that the pesticide DDT had on birds of prey during the past century.

That the barn owl will readily occupy nesting boxes bodes well for its future when suitable nesting habitat is not available. Integrating the owl's populations into agricultural settings to control rodents is also picking up momentum as the benefits of doing so are being documented and the knowledge is shared. Providing information to the public regarding the owl's services to humankind will go a long way in protecting and stewarding the future for this species.

VITAL STATISTICS

The first few years of a barn owl's life are difficult ones. It can live to be ten years old, but the chances for such long-term survival are pretty slim. One study that involved banding 572 owls found that the birds that were recovered had a mean age at death of twenty-one months. European studies of this species found similar results, with mortality rates at 75 percent in the bird's first year of life. On the other hand, one record of a wild barn owl living for thirty-four years suggests that, when conditions are ideal, a very long life for this species is possible.

Length: 14–20 inches (35.5–51 centimeters)
Wing span: 43–47 inches (1.1–1.2 meters)
Weight: 17 ounces (482 grams)

Eastern Screech Owl (*Megascops asio*)

I spotted its splayed form on a roadside snow bank caught in the lantern light atop our bobsled as we skidded along behind a team of horses through a winter night in Michigan. A romantic memory formed as an eight-year-old and was made all the more enduring by the experience of stopping the team and running back to pick up and feel the soft and pliable form of an owl for the first time. The concept of what subtle beauty can be was born in me as I examined the bird under the lights in my uncle's kitchen. With those varied patterns of black, gray, and white across its chest, wings, back, and head, the owl didn't require a lot of color to dazzle my nascent artistic side with its plumage. I still recall the sharp-hooked talons that snagged my fingers and sent me wondering what it must catch with its feathered toes, and how it did so. I later took a correspondence course in taxidermy, but that

4.3 Eastern screech owlets in nesting cavity with blind snakes.

hadn't happened yet, so the best I could do to keep the owl was to cut off its wings and pack them in my suitcase for my return trip back home to Southern California. I still have them, and their pliable feathers with the sound-softening frilled edges of the outer primaries are intact and just as beautiful as they were when I first touched them more than sixty-five years ago.

RANGE AND HABITAT

There are currently five subspecies of eastern screech owls, which in general prefer a tree-dominated habitat in the broad range of landscapes they occupy in eastern North America. This adaptable owl is found from the southern edges of the boreal forest in Canada down the midsection of the continent, along the East Coast into the south, westward along river valleys of the Great Plains, and farther south into eastern Mexico.

4.4 Range map of eastern screech owl in North America. (Question marks in the maps indicate that the range in this region is unconfirmed.)

These medium-small owls, like their counterpart the western screech owl (which it interbreeds with), have readily taken up residence around people where nesting and feeding requirements are satisfied. As a nonmigratory owl, the bird can be found in its territories throughout the year.

FOOD PREFERENCES

The eastern screech owl takes a broad range of prey. One might even say that if something crawls, the owl will give it a try. Among invertebrates, it feeds on crayfish, earthworms, beetles, centipedes, cicadas, leeches, and pill bugs. The mammals it eats include mice, bats, voles, shrews, and an occasional young cottontail rabbit. It consumes reptiles and amphibians such as blind snakes, lizards, skinks, frogs, and salamanders. This owl is very much a perch-and-pounce hunter, taking its prey on land and in shallow water, catching minnows, catfish, and sunfish when possible. The birds it preys on can be small, medium, or large: tits, wrens, jays, pigeons, pheasants, and even one recorded nine-pound chicken (four kilograms), killed and eaten. Captive owls in this species consume up to one-third of their body weight per night.

VOCALIZATIONS

Descending trill and whinny calls are used by the male eastern screech owl to establish and defend his territory. He does this with his lower-pitched voice.

A trill call is also used in the male's advertising of a potential nest to a female and serves as a contact song among family members.

A hoot is used to warn of a potential predator's approach and as an expression of alarm. Bark calls are a step up from the hoot and mean the owl is alarmed. A screech, the namesake of the species, is uttered when the owl is agitated and is defending its nest.

Bill-snapping is another expression of the owl's agitated condition, and one would think that in all owls it is used to remind the agitator that the owl has a powerful biting beak.

COURTSHIP AND NESTING

Monogamous for the most part, the birds pair up by late winter, with the male taking the initiative in showing the female nesting sites for her to choose from. The male's solicitations are sweetened with a gift of food that also is an indication of his fitness and his capacity to supply the female with meals throughout the egg and hatchling brooding periods.

These owls readily nest in natural tree cavities with small entries, but do have a preference for flicker nests of previous years. The female selects the site as early as January and as late as June, with four to six eggs laid one after the other, a day or so apart. Incubation begins immediately, which of course results in asynchronous hatching with appreciable size differences in the young birds as they begin to grow. Like the western screech owl, this species' requirement of a fairly precise diameter at the entry to the nest ensures that the owl fits snugly, but the heads of marauding opossums and raccoons are blocked.

Variations exist among the several subspecies of this owl; a few include the onset of courtship, nest site selection, clutch size, and number of days of incubation before hatching. On average, however, the eggs of this owl begin to hatch after twenty-nine to thirty-one days of incubation by the female. Throughout this period the male provides for the female, and, once the owlets have hatched, this commitment continues. Her mate will provide all the prey, because only the female will remain in the nest to brood the youngsters for as long as two weeks.

By around their fourth week, the nestlings become fledglings and leave their cavities to hop, flutter, and step into the larger world, but they won't be capable of flying for another several days, and even then they will lack strength. If grounded, they are capable climbers, pulling with beak and moving foot over foot to regain some elevation in a bush or a tree. In this outside world the adult pair will feed them for another eight to ten weeks. Of the average brood leaving the nest, only 36 percent of the birds will survive their first year.

Dr. Fred Gehlbach, the authority on eastern screech owls as well as other species, noted an unusual symbiotic relationship between these owls and blind snakes. Some of the snakes captured as prey for the young were left alive after being transported to the nest, where they consumed maggots and actually provided something of a clean-up service to the family of owls. Likewise, acrobatic ants were in the nests and of no harm to the young, although they too provide hygienic benefits by eating the leftovers from the birds' meals and, with their biting propensity, serving as a discouraging presence to any predator that might seek to enter the cavity.

THREATS AND CONSERVATION

The eastern screech owl does fall prey to larger owls and woodland raiders, such as the aforementioned raccoons and opossums. Taking up residency in company with human populations brings on the dangers of feral and domestic cats, which have a major impact on all bird populations, including owls.

The young owls in particular collide with vehicles, and both young and adults will fly into windows. Placing mesh fences in former open flyways will injure the birds as well. Still, the greatest impact remains the compromising or destruction of their contiguous woodland habitats where nesting, roosting, and foraging are provided.

Gehlbach's experience with this species is particularly insightful, because he provides recommendations for not simply sustaining their populations but also for restoring and introducing the birds to new and appropriate locations. Providing nesting box details and advice, he encourages responsible individuals to take up the cause of supporting the species, which can thrive in our neighborhoods and readily and successfully feed on insects attracted to our street and patio lights.

VITAL STATISTICS

A captive eastern screech owl lived to be fourteen years old.

Length: 7–10 inches (18–24 centimeters)
Wingspan: 18–24 inches (46–61 centimeters)
Weight: 3–8 ounces (85–226 grams)

Western Screech Owl (*Megascops kennicottii*)

Most naturalists can conjure up a memory or two about a particular wild animal that lit the first fires of their interest in nature. There is no doubt that spending time with western screech owls as a child had a major effect on how I began to respond to creatures I would find in the woods, fields, and along the shores of my home in southern California.

I was eight years old when my parents moved farther out into the edges of the San Fernando Valley, where there were still orchards below the oak- and chaparral-covered foothills. Tributaries of the Los Angeles River had yet to be encased by sterile boundaries of concrete, so there were unlimited paths of exploring adventure available. These were the days before television had much of a grip on the attention of a kid, so my companions and I were more participants with what we were discovering than passive and vicarious observers. I believe our senses back then were more keenly attuned to what was going on around us, because there were fewer distractions or alternatives for our time and energy. We cut our teeth in the real world.

Late one fall afternoon I caught a glimpse of some flying form crossing our yard to perch in a sycamore tree adjacent to our home. Then, there always seemed to be no alternative but to take a look and see what it was, and in a minute I was standing below the first owl I had ever seen outside a zoo. From not more than five feet away the diminutive bird stared down at me with what seemed to be a fearless intensity as I likewise studied him. (I say him, because this memorable encounter set me off investigating, and I soon learned it was a western screech owl, and inasmuch as males are smaller than females I judged this bird to be a male.)

4.5 Pair of western screech owls.

A childhood memory can expand over time, but I do recall this owl routinely descended from its roost in a nearby pine to land on that same branch in that sycamore, evening after evening throughout the balance of the year, and I would often be there to meet him. I became convinced that the bird not only recognized me but also favored starting his nocturnal rounds with our visit. Before he pitched from his perch to fly off silently, I would whisper a few questions to my friend: "Where are you going each night? What do you eat? Do you have a nest and where? How do you fly so silently? How can you see so well in darkness?" And while I never got direct answers, my one-sided conversation provided a basis for interest and investigation throughout my life.

I grew up in the 1940s, and I soon discovered that this species was common in the valley, because the extant stands of riverside trees, scattered groves of walnut and orange, and even the big planted palm trees along our streets suited the birds' feeding and breeding requirements. I recall the day my mother brought home an orphan

4.6 Owl and child contemplate each other.

owlet she had picked up on the school playground where she worked. Such was the presence of these owls at the time that we put the youngster out on our porch the following night, and it was immediately adopted by our resident parent owls, which had young of about the same age and began feeding it.

RANGE AND HABITAT

The western screech owl is a bird of western North America that occupies a variety of woodland and forest habitats and favors low-elevation riparian deciduous woodlands. Their range typically includes coastal forests from Alaska down into Baja California and eastward in the contiguous states as far as Montana, Colorado, New Mexico, and west Texas and southward into the highland forests of Mexico.

Diverse in their choices of residency, these owls are nevertheless most often found in deciduous forests adjacent to wetlands, rivers, and streams. In their

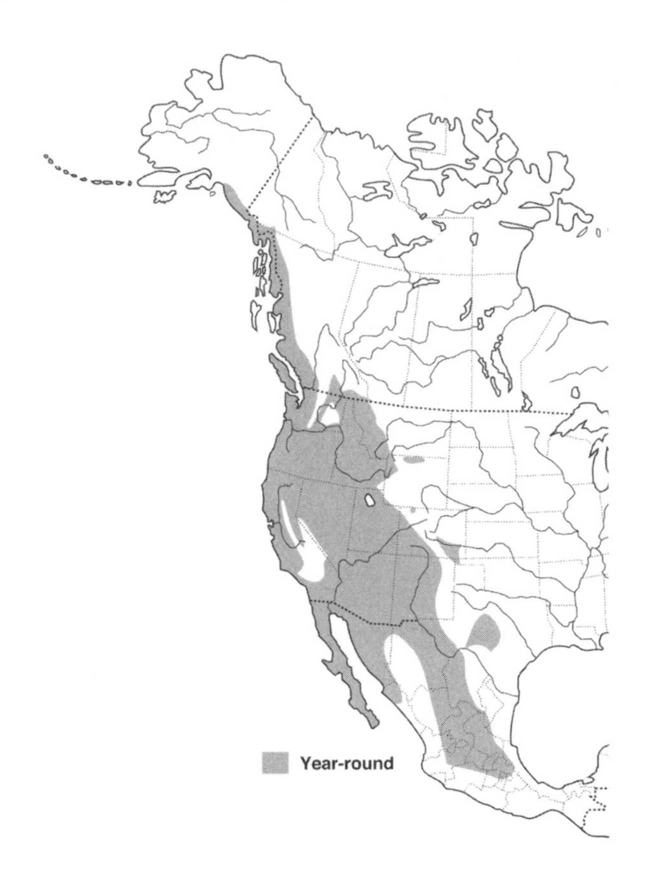

4.7 Range map of western screech owl in North America.

Sonoran desert populations they occupy mesquite riparian zones, and although they reside in a range of forested habitats, the presence of water also seems to be a determining factor in attracting them. That there are forests, whether composed of cacti or coniferous or deciduous trees, is critical to this cavity-nesting species.

FOOD PREFERENCES

A very eclectic feeder is this owl! Small but powerful and big-footed, they take prey as small as caddis fly larvae and carpenter ants or as large as a cotton-tailed rabbit. This diversity of taste results in a varied diet that the bird may have both from year to year and from place to place. Small mammals predominate, however, and include wood rats, kangaroo rats, pocket mice, deer mice, and shrews. Birds are also a favorite if available, including northern flickers, robins, Swainson's thrush, violet green swallows, and house sparrows. Depending on the owl's location, aquatic animals can dominate, and one family may well subsist largely on crayfish while another along the coast catches tidepool sculpins. Early studies of this species by Charles Bendire showed the owls were catching trout and mountain whitefish. In the southwestern part of its range the western screech owl catches scorpions and feeds intensively on the Jerusalem cricket. During the winter in Idaho small mammals and birds are sometimes beheaded and stored for later consumption.

VOCALIZATIONS

Several different calls are distinctive in this owl's assortment of vocalizations. The "bouncing ball" is a series, averaging ten in number, of soft whistling hoots that quickly become more closely spaced. This is an advertisement or territorial call often given in association with courtship and selection of nesting site.

Both male and female owls produce a sharp scolding call best described as a bark or cluck. This exclamation is produced if a potential predator becomes active around the pair's nest site.

A whinny is given by nestlings, fledglings, and adult females. It is clearly a begging-to-be-fed call, and if the male fails to feed the begging birds, the call can pick up its rate and intensity through the day or night and reach what might be described as a pitch of pleading for food.

Chirping is a multipurpose exclamation given by both sexes and serves to express alarm, assemble the fledglings for roosting, and announce the arrival of a food-bearing adult. There is some indication that it is given by the adults from the branches surrounding the nesting cavity in an effort to coax a fledgling from its interior.

COURTSHIP AND NESTING

Throughout their range, the male owls begin advertising their territories as early as late January through February, and within a few weeks higher-pitched responses to the males' bouncing-ball calls are given by the female. By the end of March and into early April, courtship has intensified with dueting calls, mutual preening, and the male's gifts of prey to the female in and around the prospective nesting site. Intermittently during this period mating occurs.

Preferring a cavity excavated by a flicker or a pileated woodpecker, the female eventually selects the site suitable to lay her clutch of two to seven eggs, which begin hatching on average after thirty days of incubation. Throughout this period and over the next three weeks after hatching, the female and the young depend solely on the male to provide their food. By their fourth week, the young are clamoring for glimpses of the outside world from their nesting cavity. By this time the female leaves the nest to join the male in capturing prey and feeding the young. Fledging of the increasingly active owlets begins shortly after this time.

Once the fledglings have awkwardly fluttered out from their nest site, they remain nearby for several days, as their flight skills quickly increase and allow a greater range of movement. They remain with the family group for at least another month, while the juvenile birds continue to be fed by the parents and develop their own skills at securing sustenance. By their fifth week after fledging, they are more independent and begin to disperse into the larger environment.

THREATS AND CONSERVATION

Larger owls and hawks prey on the western screech owl, particularly the recently fledged youngsters. Spotted, barred, and great horned owls capture them, as do Cooper's hawks. Raccoons climb to the nesting cavities intent on snagging a young bird from a cavity, and even a gopher snake is recorded as capturing an adult bird. Where these species have adapted to suburban environments, domestic cats and dogs kill grounded fledgling owls.

A juvenile owl that has made it through the first few months of life in the company of its supportive parents faces the ever present threat of starvation as prey availability diminishes with the approach of the fall and winter months.

Another threat to the young inexperienced owls comes from collisions with automobiles as they innocently hunt the borders of local highways. Every year the bodies of first-year birds are found along roadsides. Like any bird intent on getting from one point to another in rural or suburban settings, these owls are prone to collide with large uncurtained windows, and more than a few have been brought to veterinarians stunned if not seriously injured from such encounters.

There has been some evidence that pesticides have transferred into the diet of western screech owls, and, as in the historical case of the peregrine falcon, thinned the eggshells of these birds. The versatile and opportunistic prey choices these owls make, from small mammals and birds to beetles and carpenter ants, expose them to the poisons humans apply to limit the numbers of these pests.

The most serious threat to this species' welfare, however, is loss of habitat. The availability of riparian forested areas for hunting, with older trees possessing woodpecker cavities for breeding, is imperative for the owl's well-being. In the southwestern portions of its range, this is particularly so, as much human development (housing, agriculture, or industry) has greatly diminished the conditions essential to the owl.

Whether for development or commercial purposes, the judicious cutting of trees and silviculture with owls in mind will benefit the birds. Along with this practice, inasmuch as these owls are very adaptable in accepting nesting boxes when flicker cavities are unavailable, an effort to provide such accommodations will certainly be to their advantage.

VITAL STATISTICS

A pair of captive birds in Washington state lived to be nineteen years old, and a banded wild bird recovered in Claremont, California, was aged at thirteen years. With as many as perhaps ten subspecies, this owl has considerable variation in size, but the averages are as follows.

Length: 7.5–11 inches (19–27.5 centimeters)
Wingspan: 18–24 inches (46–61 centimeters)
Weight: 5 ounces (142 grams)

Northern Saw-whet Owl (*Aegolius acadicus*)

Superficially, the tiny owl appeared to be little more than a handful of broken and disheveled feathers. Found along a roadside, it had likely been hit by a car and stunned to the degree that it was unable to fly, so there it remained where it was soaked for God knows how long by an unrelenting rainstorm.

What life remained to the northern saw-whet was probably advanced by whatever adrenaline it could muster to put up a feeble fight when picked up. I knew that the bird, soaked to the skin, had been severely chilled, and that hypothermia could easily kill it within a few hours, so placing it on a heating pad in a warm enclosure might be its best hope for recovery. From the moment I picked up the little owl and all during the transfer to a warm, dry retreat, its wide-eyed fiery glare was unrelenting, as was the firm grip it kept with its tiny toes on my finger.

The warmth of the pad and softness of the towels in the box seemed to relax the bird, and when I placed a towel over the top of the enclosure to darken it, it was intended to give the bird a break from the trauma of handling before I sought to feed

4.8 Northern saw-whet owl.

it. The bird's energy reserves were exhausted, but in advance of picking it up again I thought it best to bring its diminished body temperature up to normal.

I gave the owl an undisturbed two hours in the improvised incubator. Opening the lid, I was relieved to find that even over that short time its sodden feathers had been dried and roused enough by the bird for it to begin to take the shape of an owl again. It was imperative, however, that I get some liquid and somewhat solid food into it if possible. Regardless of how supportive the heat of the incubator might be, if the bird did not have sufficient nutrients and fluids in its system it could easily succumb to dehydration.

Feeding any wild adult altricial (naked, blind, dependent on the parents for food) bird species that has advanced well beyond the innate gape and gulp period of its infancy is a problem. Holding in one hand the now warmed up and slightly restored owl, I offered a very moist and tiny morsel of raw chicken thigh. My hope was that the bird would bite at it and then reflexively swallow it. No luck. All I got in response to my waving the meat back and forth was a fixed stare and a beak clamped shut.

Over the next half hour I pinched the lower edges of the owl's beak to get it to open to a point that I could slip a sliver of food into its mouth. Once the food was in its beak, I held the beak shut so the bird did not shake its head and flick it out. Eventually the fluids in the morsel triggered the owl's swallowing reflex, and down it went. After another half dozen feeding sessions, the owl was taking the food directly from my fingers with an increasing, appreciative energy.

Two weeks of care and feeding combined with evenings of flight about the living room appeared to restore this northern saw-whet owl. Although its full-plumaged size and weight suggested it was probably a female, hunting kills were never tested in the house; at the time of her release our woods were full of insects and harboring plenty of mice. I felt confident about her future when she again took to the wild woods.

RANGE AND HABITAT

Found throughout most of North America, this smaller owl occupies both deciduous and coniferous woodlands. Its forest residency can occur in elevations ranging from sea level up to 10,500 feet (3,200 meters). For foraging, nesting, and roosting a preference is shown for the dense cover of a middle canopy. In the Idaho portions of its range, however, it avoids the dense forests to occupy riparian and savannah country.

For obvious reasons, the movements of nocturnal owls are yet to be fully understood, but it is certain that some populations of this species do migrate under cover of darkness in the fall and spring. A September movement in an eastern population is known to follow a route along the Atlantic coast and somewhat into the interior of the continent from Nova Scotia to North Carolina. The following spring, beginning in March and continuing into May, the birds begin their return to the north. Other groups of northern saw-whets follow a migration route in the central portions of the continent from Ontario down through the Ohio Valley and into Kentucky.

FOOD PREFERENCES

Throughout its range, mice are part of the preferred diet, along with voles, bog lemmings, and pocket gophers. To a lesser degree saw-whet owls consume birds, particularly passerines that are caught during migration. Some of these include kinglets, robins, and even birds as large as rock doves. An unusual report of a northern saw-whet eating a northern pygmy owl leaves much to one's imagination as to what kind of a battle occurred to enable one small owl to subdue the other.

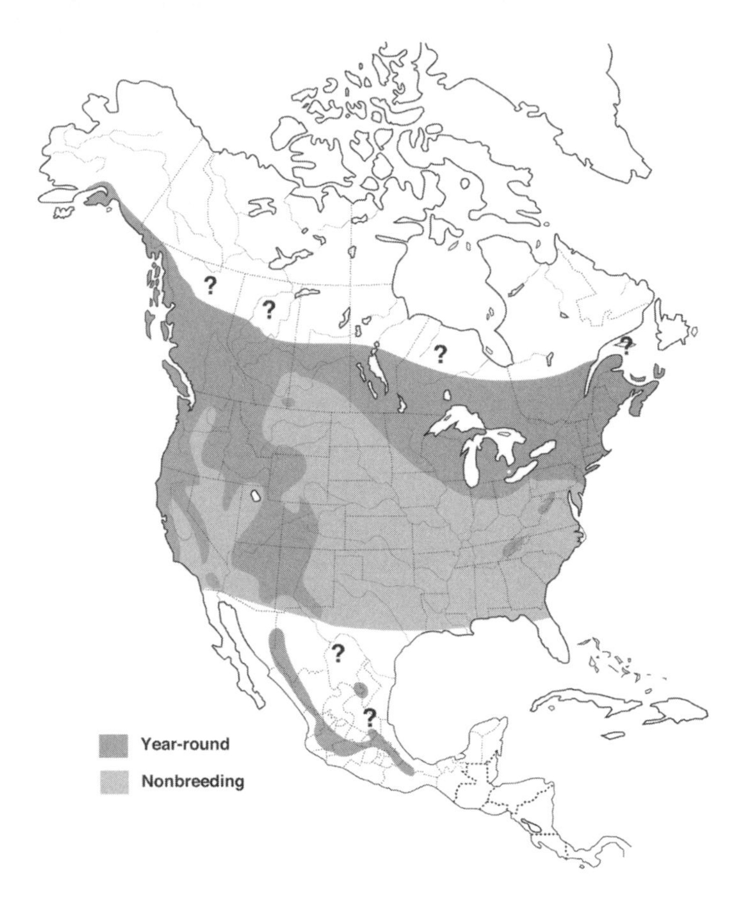

4.9 Range map of northern saw-whet owl in North America.

Insects are a large part of this owl's diet, including grasshoppers, spiders, and beetles. Along the coasts where this bird is an occupant, intertidal invertebrates are eaten.

VOCALIZATIONS

Saw-whet owls are very nocturnal, and they give their advertising calls from thick cover, where they remain relatively undetected. Such a strategy is very important in a forest where predators will attempt to locate them. The species begins calling a half hour before sunset and may continue through the night until shortly before sunrise. Committed to his task, one male northern saw-whet called continuously during these times for a span of ninety-three days.

Whistle notes, delivered in a series and at a constant pitch, are typically a male's advertising call. The call can be heard in the forest at nearly a thousand feet (three hundred meters) and carries over water more than half a mile (as far as a kilometer).

A versatile singer, this owl uses at least nine different identified vocalizations.

A soft whistle, similar to the advertising call but at a lower intensity, is given by the male to announce his arrival with food for the female at the nest.

Squeaks are a loud, sharp sound, not unlike that produced when sharpening a saw blade, hence the owl's name, saw-whet. This call is delivered during breeding season and likely to be a territorial declaration.

"Tisst" is given by the female in response to a male's advertising song and when she solicits food from her mate.

Chirrups are begging calls by youngsters that morph into "Tsshk" as the young develop.

"Chuck" appears to be an expression of disgruntlement or dismissiveness, expressed when a captured bird is released.

COURTSHIP AND NESTING

Male northern saw-whets are advertising with song as early as late winter on the territories they may have maintained throughout the year. Although generally monogamous, in years of abundant prey a male owl may have more than one mate and may successfully fledge as many as eleven young from two nests. Not to be left out of such an arrangement, females occasionally leave the fledglings, re-mate, and raise another brood.

Five to six eggs are laid from late February to early April in a nest site chosen by the female from those her mate has offered up for consideration. The site is typically a northern flicker or pileated woodpecker cavity from the previous year. In parts of their range where nesting cavities and prey are both abundant, the owls may share their territory with northern pygmy and boreal owls.

After twenty-seven to twenty-nine days of the female's exclusive incubation, the young are hatched. Throughout this time the male has been the sole source of sustenance for his brooding mate and will continue to play this role for the next two and a half weeks for the entire family. Around this time the female relinquishes her brooding duties and leaves the cavity. The male continues to assume the greater share of feeding the young and appears to get little or no help from his mate.

Thirty or so days from hatching the young fledge. Once they do, these owls have some flying ability, rather than being limited to just gliding or climbing about, as are screech and flammulated owlets when fledging. Over the first month out of the nest, the male still provides most of the food for the young.

THREATS AND CONSERVATION

The northern saw-whet faces challenges from the resident predators that occupy the forests where the bird resides. Cooper's hawks, barred owls, and great horned owls frequently capture northern saw-whets.

A greater threat for these birds is starvation, which is a common cause of death for this species in its first year of life. Even before they have left the nest, however, the youngsters can be infested with fleas and flies, leaving the birds severely weakened and even more susceptible to starvation or disease.

The young owls and adults occasionally are attracted to the borders of roadways where small mammals are sometimes abundant. A bird perched in the trees adjacent to the highway sees a rodent flushed by the sound of an approaching car. Its pursuit response is triggered, and its flight trajectory and that of its racing prey often intersect with the path of the speeding automobile. The inevitable collisions with roadway traffic take the lives of many of these owls every year.

As with other species of owls, expansion of human development has a subtle but significant effect on the northern saw-whet's populations. Breeding, roosting, and hunting environments for the owl are degraded and compromised as woodlands are cut for commercial, residential, industrial, and agricultural interests.

All of these impacts can, to some degree, be mitigated. Where natural nesting locations are few or absent, nesting boxes are readily used by northern saw-whets. Set-asides of woodlands in land-use planning will also ensure the presence of these and other owl species. A concurrent public information and education effort that relates to the virtues and benefits of woodlands can go a long way to encourage an informed public to take pride in these areas and an interest in owls.

VITAL STATISTICS

Richard Canning's work with this species is particularly revealing. Of seventy-five birds he banded for a study, only half survived their first year, and only six of the first-year survivors made it to the age of three. One bird of the banded group lived to age seven. Although the rigors of survival are difficult for this species, we do know

4.10 Cooper's hawk pursuing a saw-whet owl.

they can live appreciably longer under some conditions, as a captive bird was reported to survive to the age of sixteen years.

To maintain its weight, a 3-ounce (90 gram) female northern saw-whet owl in captivity would eat 0.6 ounces (17.5 grams) of fresh prey each day, or almost a fifth of its weight.

Length: 7–8.5 inches (18–21 centimeters)
Wingspan: 17–21 inches (43–51 centimeters)
Weight: 3 ounces (85 grams)

Short-eared Owl (*Asio flammeus*)

Winters along the estuaries of Washington's Salish Sea provide a stage for raptor performances. Falcons are slicing into the flocks of ducks and shorebirds as heavy-winged hawks lumber off perches to plunge into ground cover after meadow voles. Threading the needle through all this action and often traveling in company are northern harriers and short-eared owls. Convergent evolution has fashioned these two species to be similar in form and habit. Both birds turn on a dime when some hint in sound or sight requires their investigation. Both hawk and owl have disks of feathers that concentrate the light to the eye, allowing the highest visual resolution of movement or activity below them. Although reflecting the light to the eye, the facial feathers are open enough to allow the full spectrum of sound to be picked up by the bird's auditory canals as well.

In the field at play with the owls I've accidentally come upon communal roosts of these owls. Once, with my wife and a friend, we stepped into a communal roost of more than thirty birds. There was a startling and bewildering moment as the birds erupted from underfoot, going in all directions like a massive cloud of enormous blond moths. They whirled about for a few minutes and then settled back down collectively into the ground cover and disappeared. There may have been a survival strategy at play here. Such a confusing dispersal of owls would have made it difficult for a predator to get a line on a single target. I wondered if within these gatherings of owls some prenuptial getting acquainted might also occur before the breeding season ahead.

The short-eared owls I've handled have all been injured birds, most crippled by collisions with cars and one or two shot or wing-damaged from colliding with a low wire fence obscured by overgrowing grasses. Without exception they are gentle birds when in hand with a minimum of hissing and bill-snapping that other species

4.11 Short-eared owl.

offer up as protest. The owls in the field, however, are anything but passive. I've watched these birds aggressively respond to other, far heavier raptors. They trade chases with American rough-legged hawks and sometimes strike them. Snowy owls and red-tailed hawks are harassed when they trespass on what the owl considers its hunting territory.

RANGE AND HABITAT

Across the northern portions of North America, throughout Europe, across into Asia to the Bering Sea, this species is observed as widely about the world as any owl. There are even populations in South America, out onto the Galapagos and Falkland Islands. As a nonbreeding wintering species, this owl occupies the balance of the United States southward to the middle of Mexico.

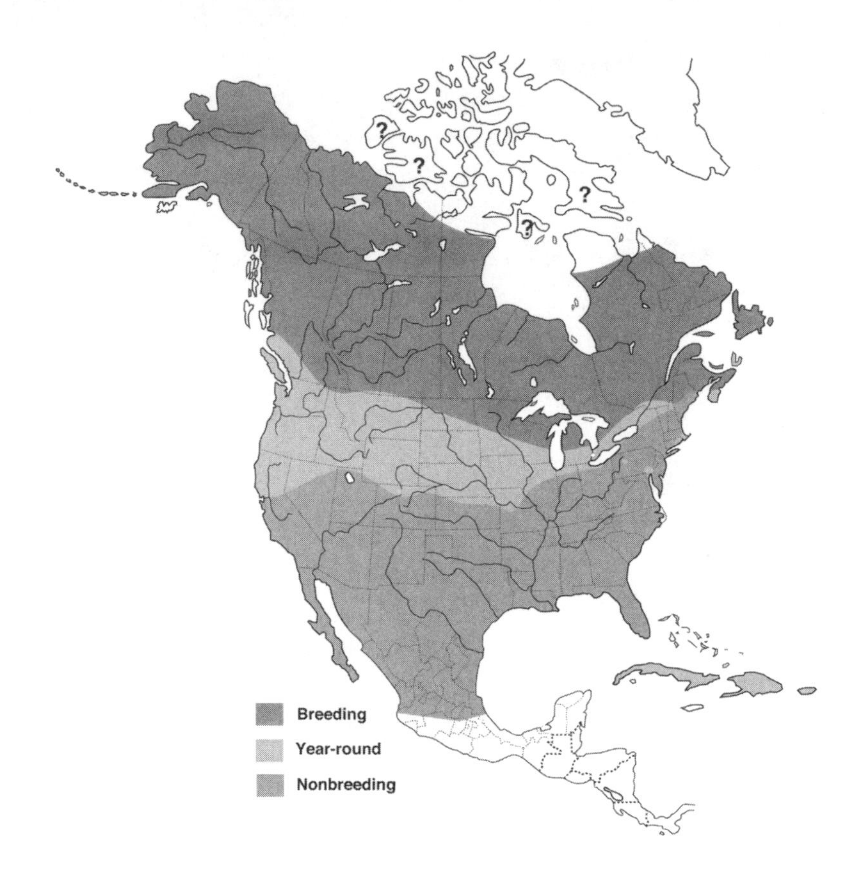

4.12 Range map of short-eared owl in North America.

Short-eared owls prefer open country, prairie, and grasslands along the coast, shrub steppe, and tundra where small mammals reside.

FOOD PREFERENCES

Small mammals dominate the diet of these owls, including voles, deer mice, and meadow mice. They will take small birds occupying the owl's habitat, such as the meadowlark, as well. Insects favored include grasshoppers, cutworms, and beetles.

VOCALIZATIONS

In threat situations the short-eared has its own distinctive bark, as do other owls. A "waak–waaa" is delivered in threatening situations or when harassing a competitor such as a northern harrier or rough-legged hawk that is hunting in its territory.

There are also assorted squeaks and whistles the intent of which is unknown, but the "keeeow" call may have a particular purpose as the birds hunt. At times they produce this call as they course back and forth over fields dense with grasses. It has been speculated that the intensity of this call is meant to startle a mammal otherwise concealed. Its movement would give away its location to the owl hovering overhead.

A "hoo-hoo-hoo-hoo" vocalization is reserved for courtship and often accompanies the owl's distinctive wing-clapping sounds, produced as the male engages in a "sky dance" courtship display before the female. The sound resembles that of slapping one's thighs in rapid succession.

UNIQUE STRATEGIES

As in some other owls, including the barn, great gray, boreal, and saw-whet owls, the short-eared is equipped with modestly asymmetrical openings to the auditory canals. This is a distinct advantage in locating invisible prey. The owl positions its head so the sounds of a scurrying mammal are perceived equally in both ears. At this point the target is directly before the hunting bird, and its plunge into the cover is precisely targeted to its prey.

Like some other owls, including its very close relative the long-eared owl, these birds have a distraction display to lure an intruder away from their nests. They will fly to the ground before a threatening animal and flap about helplessly, appearing to have a broken wing. If all goes well, they stay just beyond reach and gradually lure the predator away from their nest.

COURTSHIP AND NESTING

Unlike most male owls of other species, the short-eared does not use elevated perches to advertise territory and initiate courtship. Not to be deterred, this owl, like the northern harrier, takes to the sky to display and call a prospective mate's attention to a nest site. As the communal roosts of the owls begin to disband in late winter, male owls perform a special flight sequence over a female owl that Denver Holt and Shawne Leasure refer to as a "sky dance." With the female owl perched on the ground below, the male soars overhead as high as 500 feet (150 meters), calling as it does so. At the top of its climb the bird stoops falconlike, interrupting the plunge with a successive clapping of its wings together a dozen or more times. The male often repeats the performance before descending to the ground cover, where he is joined by the female and copulation may occur.

By April and early May, in the more northerly portions of its range, the owl has scraped out a nesting site in the dried vegetation from the previous year. Here, three to four eggs are laid, and by the third week the first egg is hatching. During incubation and most of the brooding period the male is the sole provider of food.

Thirty to thirty-six days after hatching, the owlets have developed to a point where they begin to move out from the nesting location in different directions. This manner of fledging serves as an avoidance tactic, as ground predators might find a single owlet but won't come upon the lot of them.

THREATS AND CONSERVATION

Competing for wintering habitat occasionally results in the capture of these owls by larger raptors. Migrating goshawks and snowy owls that periodically irrupt to move southward from the Arctic and occupy coastal lowlands occasionally prey on these owls. Of greater threat are opportunistic foxes and coyotes that take the eggs and young of this species. Feral cats, too, can have a serious impact on these ground-nesting owls.

As widespread as the short-eared owl is, populations of this species have diminished as their feeding and nesting habitat in open land is converted to intensive agriculture and cattle grazing. This is particularly so in the midwestern and northeastern United States. Likewise, such feeding and nesting locations are becoming unavailable to the owls in southern Idaho, portions of central Oregon, and south central Washington. Even where the owls are able to sustain their numbers in an agricultural community by eating rodents attracted there, the birds can suffer. In California, M. N. Rosen found that forty-four of these birds had died from fowl cholera passed along to the owls from the mammals they had consumed.

Invariably, while hunting the open areas crisscrossed by highways, these birds are often killed by vehicle traffic. In spite of this inevitable consequence of our expanding development, there are strategies of restoration and conservation that can sustain the welfare of this owl. Massachusetts has a Natural Heritage Program that maintains large tracts of contiguous land suitable for both nesting and hunting short-eared owls. The state has combined this commitment with a public education effort that includes both information on the birds and opportunities for monitoring them that can reduce human disturbances and predation. Furthermore, any land management that applies planned burning of grasslands for gallinaceous and waterfowl species can also benefit the short-eared owl.

4.13 Long-eared owl in flight.

VITAL STATISTICS

A banded wild short-eared owl recovered in North America lived a little over four years, and a European bird lived to be twelve years and nine months.

Length: 14 inches (37 centimeters)
Wingspan: 38–44 inches (100–110 centimeters)
Weight: 11–13.5 ounces (315–382 grams)

Long-eared Owl (*Asio otus*)

For many years, in late spring, I would routinely go to the Columbia Plateau of eastern Washington to observe the breeding activity of the abundant species of birds of prey there. Prairie and peregrine falcons had nest scrapes on the basalt cliff ledges, while red-tailed and ferruginous hawks placed their eyries of sage branches there as well. Great horned owls appropriated the abandoned nests of these hawks to raise their young, while barn owls retreated to the hollows in the stone to lay their eggs. In the canyon bottoms where tree clusters are nurtured by ground and surface water, American kestrels occupied old flicker cavities, and Swainson's, sharp-shinned, and Cooper's hawks built nests. Long-eared owls flattened the domes of old magpie residences into shapes that were suitable to brood eggs and young. At ground level and below, short-eared owls bred amid the grasses, as did northern harriers. Even below-ground something was going on, as burrowing owls expanded the tunnels of badgers and ground squirrels to their specifications and raised families.

It was not only the suitable nesting sites that encouraged these birds to breed here, but also the abundance of small mammals, reptiles, other birds, and insects that were a food formula perfect for raising young raptors. This cornucopia of food for falcons, hawks, and owls was so plentiful in some years that the edges of some of their nests were stacked with the likes of pocket gophers, voles, and wood rats. One such nest was that of the long-eared owl that I studied there, and from a distance I would only occasionally see the top of her long ears pop up above a surrounding circle of mammal carcasses.

This particular bird and her mate provided me with an unusual experience, although I would learn later that distraction strategies are not exclusive to long-eared owls. When I approached the nest two weeks after the last youngster had hatched, hoping to review the catch of prey, what I judged to be the male owl swooped from his nearby sentry perch to plunge into the ground cover a few meters before me. For a moment I thought he had sought to catch something there, but to my astonishment the bird remained on the ground, flopping this way and that as if seriously injured. High-pitched cries of distress accompanied his display, and if I closed the gap between us he awkwardly took to the air only to drop to the ground again at some distance and repeat his feigning injury. I followed the owl, and was surprised at how effective this distraction strategy was in leading me farther and farther away from the owl family. I was well away from the nest and its stand of willows when the owl appeared to recover his health and powers of flight miraculously, and it flew off gracefully and perched on the top of a distant column of basalt.

RANGE AND HABITAT

Occurring across most of North America and northern Europe, the long-eared owl hunts the drier open habitats while nesting and roosting in dense vegetation. It actively searches for prey chiefly at night and will course low over meadows, grasslands, and shrub steppe in the hunt.

FOOD PREFERENCES

Preferring small mammals, the long-eared owl takes montane, prairie, and meadow voles, deer and house mice, kangaroo rats, pocket gophers, and ground squirrels. As versatile hunters, these owls catch hoary, little brown, and pallid bats

4.14 Range map of long-eared owl in North America.

along with lizards and snakes. In spite of their moderate size, these owls are capable of catching long-tailed weasels and birds as large as ruffed grouse.

VOCALIZATIONS

"Hoo—hoo-hoo" is a male's advertising call, given every two to four seconds.

Females and youngsters use a mewing call to beg food from the male.

A bark or "ooack" call is uttered when the owl is alarmed or an intruder attacks.

A shriek accompanies a distraction display as the owl lures an intruder from the nest.

The clapping of wings, up to twenty times, by the male owl occurs during a courtship flight.

COURTSHIP AND NESTING

Males begin to advertise their territories with song and flights shortly after sunset as early as February and will continue into May. Their courtship displays are quite dramatic, as the owl deftly and swiftly flies about the stand of trees wherein a potential nest is located while taking deep wing beats between glides and clapping their wings together. The courtship reaches a crescendo as the male, having completed his vocal and aerial displays, descends to a perch, whereupon he begins to sway while raising and lowering his wings. In response the female, with drooping wings, crouches either on a branch or on the ground and presents herself to her mate. The male may then fly to the female, and with both birds extending their wings, he will mount her for a three-second contact. Allopreening may occur both before and after copulation.

By mid-March, the male's advertising of nest sites is finally resolved by the female making a choice. Depending on the owl's location, a typical choice for laying eggs might be a crow, magpie, or Cooper's hawk nest. Over a span of twenty-six to twenty-eight days, the female incubates the five to seven eggs; the male feeds her over that entire time.

Within three weeks of hatching, the fast-growing youngsters are out on the edges of their nest, and by their thirty-fifth day they are capable of taking short flights. The family remains together as both adults continue to feed the young for as long as eight weeks after fledging. In one study the male alone continued to feed the young for a full eleven weeks after the birds had fledged.

THREATS AND CONSERVATION

In North America, young and old birds are particularly vulnerable in the open country they occupy, because the aggressive great horned and barred owls include this species in their diets. Cooper's, red-tailed, and red-shouldered hawks and golden eagles not infrequently catch the long-eared owls. At the nest, foraging raccoons will attack the young.

As with other predatory species the ebb and flow of the food supply are always a challenge; for the young of a given year, starvation is a possibility. A causal phenomenon in the long-eared owl's mortality, as with all owl species, is the loss of suitable habitat driven by development. When riparian woodlands and tree stands adjacent to open meadows and fields are displaced by agricultural fields, roadways, shopping malls, and the like, the owl's numbers plummet.

To offset such losses in Britain, it appears that these birds will use baskets and open fronted boxes, a viable strategy in North America as well. Tree groves can be conserved and stewarded to provide for the owls' nesting, roosting, and hunting requirements.

As with other species of owls, a simple and thoughtful educational program can be developed around the presence of the birds, allowing the public to begin to develop an understanding and appreciation of the wild and intriguing company they keep in their neighborhoods. Such an informed public can provide support for the political and financial action that may be required to expand and sustain the interests of the owls.

VITAL STATISTICS

A North American long-eared owl lived to be nine years old, and one in Europe lived for twenty-seven years.

Length: 13–16 inches (35–40 centimeters)
Wingspan: 36–42 inches (91–106 centimeters)
Weight: 9–10 ounces (255–283 grams)

Barred Owl (*Strix varia*)

Over the years injured owls have been dropped off at my home in the hope that I might rehabilitate them. Most of them were found on roadsides having suffered injuries from colliding with a vehicle. Of those that survived, many were injured so badly that they would never survive in the wild. Some of these birds remained with me for the rest of their lives. Such was the case of a sturdy male barred owl with a wing so badly shattered that setting it was impossible, and all I could do was trim the feathers so he might hop from branch to branch more comfortably. We named him Buttons, because his dark eyes reminded us of polished fasteners on my wife's coat. He would keep our family company for more than a decade.

Beginning with his first week in his spacious aviary this robust bird began, throughout the night, a series of proclamations declaring his presence. He regularly produced a continuous series of hoots, shouts, screams, coughs, and gargles, interspersed with the species' more familiar "who-who-who—whocooksforyouall." It wasn't long before the neighbors began to make cautious inquiries as to what manner of beasts we were harboring behind our hedges.

4.15 Barred owl with captured western screech owl.

Perhaps his most remarkable exclamations were offered up during my summer wedding held here at my home. The event was outside and Buttons's enclosure was immediately adjacent to where the sixty guests were seated. At about midpoint in the ceremony and after an unsteady flute solo, the owl, without missing a beat, began to loudly offer up his musical contribution to the event. We were treated to several minutes of "whoop-whoop-who—who" and several screams before he settled down and the ceremony was completed. Somehow it all seemed very appropriate.

Buttons was a regular source of artistic inspiration to me, with his vast range of postures and displays. As the owl's trust in me grew, I could study him close at hand inside his aviary. Once there, Buttons would edge over to me along his perch and solicit a head scratching, and I in turn would offer my scant plumage, which he would preen about my ears. Occasionally, Buttons would catch an unsuspecting rat. The rodent was seeking food scraps he left, and the bird would pounce on it. If I approached him on these occasions, he would crouch down over his prey and turn so he might spread his good wing over it to conceal what he had. This posture reminded me of a strategy employed by one of our daughters toward a sister when a particular treat was in short supply.

Supplying injured owls with food is an ongoing commitment, and I had gotten very good at spotting fresh edible road kills when going about town. Eastern gray squirrels were a frequent source of owl chow, and one pickup was particularly memorable. Headed for a meeting and dressed in a suit and tie, I pulled my car off to the side of the road and walked back into a residential area where I had seen an animal flattened on the asphalt. Although a fresh fatality, the squirrel was something of a pancake, having been run over a couple of times, so I had to bend down and peel it off the surface. When I looked up I met the fixed stare of an older lady watering her garden alongside me. I felt compelled to say something, and thinking of my hungry owl, I held the squirrel up in front of me and said, "Lunch." She nodded knowingly and went back to her watering, leaving me with the feeling that my retrieving the roadkill was no big deal to her. She must have thought I was going to cook the mammal for a meal, and I wondered later if she had a few recipes to share.

RANGE AND HABITAT

Up until the last three-quarters of a century, the barred owl's range was almost exclusively eastern North America. Having a long association with an expanding human population in the East, barreds have adapted to changes in the landscape and expanded their numbers accordingly. Over the past three-quarters of a century the birds have moved westward across southern Canada and into the Pacific Northwest.

This is a forest-loving species, whether tree stands in swamps, along creeks and streams, or drier uplands. Like its close relative the spotted owl, it prefers contiguous forests with a mixture of deciduous and coniferous trees. Unlike the spotted owl, however, it is not dependent on undisturbed old growth stands for its survival.

FOOD PREFERENCES

As opportunistic feeders, these large, strong owls have a diet composed largely of assorted mammals from voles up through rabbits and squirrels. When circumstances allow, they will take to wading into water to catch frogs, fish, and crayfish. Birds such as grouse and smaller owls, particularly the western screech owl, are also consumed. Unlike its counterpart the eastern screech owl, the western screech has not coevolved with the barred owl, so the smaller owl has not had time to develop evasive strategies and behaviors around its larger relative, and is easily caught.

4.16 Range map of barred owl in North America.

VOCALIZATIONS

The barred owl is perhaps the most vocal of all the owls in North America, producing its array of booming calls both day and night. The "who-who—whocooks-foryouall" is the bird's trademark territorial advertisement and is given by both male and female.

A "hoohoohooaw" is also a common call of both sexes, and may function as a contact message between a pair of owls.

Hoots, cackles, caws, and gargles are produced when pairs of birds are engaged in spirited and raucous dueting sessions.

Bill-snapping, as with other owls, is produced when a bird is threatened or stressed.

COURTSHIP AND NESTING

After a courtship that begins even before the first of a new year, the female selects a site offered by the male, which may be the top of a broken snag, a tree hollow, or an abandoned squirrel or hawk nest. The two to four eggs require an incubation period that lasts from twenty-eight to thirty-three days, and by their fifth week from hatching the owlets are leaving the nest site.

Once fledged, the young owls will sometimes remain in the company of their parents and continue to be fed for as long as four to five months. Once their survival skills are ensured and dispersal of the young occurs, the birds, by nature, wander widely. One barred that was banded in Nova Scotia traveled in its first year nearly a thousand miles (sixteen hundred kilometers) into Ontario, Canada.

THREATS AND CONSERVATION

Like the barn owl, the barred appears to be holding its own in the company of people. In some parts of its range, however, there are local predators, including great horned owls, raccoons, pine martens, and fishers that can reach their nesting retreats and take both adults and young.

In the late summer, the bodies of recently fledged barred owls are frequently seen along highways. Going there to hunt small mammals, they have no experience with the dangers of automobile traffic and are sometimes hit. On one ten-mile stretch of highway across the South, from Florida into Mississippi, I counted eight dead owls.

VITAL STATISTICS

My injured barred owl lived for twelve years, which is twice the life expectancy for wild birds of this species. Nevertheless, one wild barred owl lived to an age of eighteen years.

Length: 17–24 inches (43–61 centimeters)
Wingspan: 40–50 inches (100–125 centimeters)
Weight: 1.4–1.8 pounds (612–676 grams)

4.17 Great horned owl with eastern screech owl prey.

Great Horned Owl (*Bubo virginianus*)

There was a time in my early life when friends claimed I covered more miles backing up my car to pick up dead birds along the highway than I did driving forward to a destination. There's some truth to this, particularly in retrieving, skinning, and studying owls. Such days are long gone, however, as there are now laws prohibiting and controlling the salvaging of wild animals.

The owls I was able to salvage were certainly put to good use, and the great horned owl in particular has been the subject of innumerable drawings, paintings, and sculptures, not to mention being a valuable teaching tool. On one occasion I was instructing a set of art classes at my daughters' school. My intent was to demonstrate how one might take a subject from nature and use it as a reference for an art project.

The owl itself immediately provoked awe for kids whose only experience with such an animal had been through children's books, television, film, or video. The recently killed big bird had been picked up from our neighborhood roadway. It was an impressive sight, with its nearly five foot wingspan, enormous taloned feet, hooked beak, and a sliver of golden glow still remaining in its enormous eyes. The two dozen middle school kids all crowded around the subject as I discussed the niche it occupied in our neighborhoods. Touching the velvety plumage and the tips of those sharp talons registered some of the reality of how these birds can fly with stealth in near darkness and capture mammals as large as house cats.

When I began to skin the bird, suggesting that an artist should understand the inside of a subject as well as the outside, the class immediately divided into the "Oh yeah, this is gonna be fun" and "Oh, no, I think I'm gonna be sick" groups, but I proceeded anyway. The revelation of the massive muscles of the bird's chest and upper and lower portions of his legs gave the youngsters an immediate appreciation of how a bird propels itself, and fastens on to and subdues its prey. When the owl's head was skinned, the enormous eyes were exposed where they extended beyond the sockets of the bird's skull. The short discussion that followed regarding the hows and whys of the bird's eye design possessed a vitality that would have been lacking without the real thing before them.

Once the owl's body was separated from the skin I told the kids, who had held fast to the front-row seats, that I wanted to see what the bird had been eating. With the accompanying sounds and smells of belching intestines, I opened up the abdominal cavity and reached up to remove the gizzard. I also discovered, too, that the owl was a male and showed them the testes, remarking that had it been a female we would have found the ovaries along the upper edges of the cavity. Holding the owl's distended gizzard in my hand, I could tell it had recently swallowed a rodent large enough to be a rat. I waited until they had a chance to make a few guesses as to what it might be before I reached inside and pulled out an entire brown rat by its tail. Gasps came from the back row, but the burgeoning physiologists, biologists, and surgeons remained near the dissection, taking it all in.

The science and art project was completed with the students eventually developing a series of drawings of a subject from nature, then fashioning a plaster sculpture and carving in the details. I did likewise with the owl and made a sketch of the

4.18 Range map of great horned owl in North America.

bird's handsome face and used it as reference for the bust of the bird I made in plaster. Later, taking a mold from the plaster owl, I made a bronze casting of the owl and took the finished piece back to show the students and follow up on some of their work as well.

RANGE AND HABITAT

This species enjoys the widest range of any North American owl, with as many as sixteen subspecies found across the continent from the Arctic to Mexico and Central America.

FOOD PREFERENCES

An eclectic feeder, the owl favors mammals from mice, rats, ground squirrels, rabbits, and hares up to marmots, porcupines, skunks, raccoons, and house cats. Birds are included in its diet, ranging in size from small owls to larger birds like coots, geese, and herons. One study described the owl pursuing a heron with such determination that it would fly back and forth three miles or more (four to six kilometers) to their colony in order to feed.

VOCALIZATIONS

Its "who-hoo—hooo" is the definitive call of the great horned owl and is usually given with the startling white of its feathered throat area in full view from a prominent perch as both advertising and declaring its territory. The resident pair will often duet using this call.

Screams are typically food-begging calls from hungry owlets.

Barking calls are uttered by adults when disturbed, and there is also a collection of shrieks, hisses, and cooing notes that have yet to reveal their full meaning.

Bill-snapping, as in other owl species, is produced when the owl is irritated or distressed.

COURTSHIP AND NESTING

These owls are monogamous for up to five years, and pairs will occupy their territory together year round when sufficient food is available.

Their mating rituals are well described in Arthur Cleveland Bent's *Life Histories,* which includes a description of the male perched alongside the female, absorbed in bowing and hooting. With tail cocked upright, the male inflates his white bib below the beak to gain attention by bobbing up and down, providing a stark contrast with the rest of its somber plumage and the night conditions. Other owls—great gray, western screech, and burrowing among them—have light-colored bibs that may be used likewise in their courtship displays. After these preliminaries, the male cautiously approaches his mate, and they respond to one another by calling in a duet.

Then, with bibs fluffed, they rub bills and preen each other until the female, assuming a horizontal position on a branch and with tail upright, is mounted by the male with his wings extended for balance, and coition occurs. Relatively speaking, it's a brief affair of up to seven seconds, during which both birds hoot continuously.

Adaptable birds, these owls are in their nests as early as the end of the calendar year or beginning in January. Abandoned hawk nests, tree hollows, broken snags, cliffs, and caves are preferred. Egg-laying occurs as early as November and as late as May. Incubation begins immediately with the laying of the first egg, and with rare exceptions the female is the sole brooder of both the eggs and the hatchlings, which emerge on average thirty-three days after laying.

The owlets remain with their eyes closed for more than a week after hatching. They nevertheless gain mass rapidly, going from a mean birth weight of 1 ounce (28 grams) all the way up to 2.2 pounds (1,000 grams) by their twenty-fifth day. On average the brood is fully feathered and capable of some flight between the forty-fifth and forty-ninth day after hatching. Once out of the nest, the young may remain with the adults into the early fall, continuing to be fed, perfecting their flight, and developing their own hunting skills.

THREATS AND CONSERVATION

Young great horned owls are taken from their nests by marauding raccoons, and if they are found on the ground they can be killed by foxes, coyotes, and badgers. Beyond this predator threat is the possibility of starvation during cycles in the availability of prey. A study of the species in Canada found that nearly half of thirteen fledged owls were dead within five months. Survival was determined by the availability of snowshoe hares.

Close studies have revealed that infections of tuberculosis and herpes virus have killed these birds. They are also subject to the effects of pneumonia and infections resulting from blood parasites. Body parasites, including blood-extracting black flies, screwworm flies, mites, and helminths (wormlike parasites), also compromise the bird's health.

Their prey itself has also contributed to the bird's death. Great horned owls have been found fatally impaled with porcupine barbs and blinded after being saturated by skunk musk.

Of course there are the usual highway fatalities, but a far larger toll of the great horned owl is taken by more direct human actions. A bird rehabilitation facility reported that of 125 live owls treated for injury, more than half had been either shot or caught in leg-hold traps. Direct deaths of the owls have also been attributed to the smorgasbord of chemicals that are applied to food crops and then consumed by pests that are caught by owls. The birds concentrate the poisons that eventually can compromise their immune systems.

VITAL STATISTICS

This species, should it survive the first few years of life, may well enjoy the longest life span of any North American owl. Banded nestlings have lived for twenty-one to twenty-two years, and one individual banded at an unknown age reached at least twenty-eight years.

Length: 18–25 inches (43–63 centimeters)
Wingspan: 50–60 inches (125–150 centimeters)
Weight: 3–4 pounds (1.3–1.8 kilograms)

FIVE

Owls of Unique Habitat

All owls are fashioned in form and behavior to exploit the resources of their habitats, but some species are uniquely fitted to their environments, and their destinies are especially closely tied to particular conditions.

The northern spotted owl has become a cause célèbre as a result of its dependence on the remaining ancient forests of the Pacific Northwest. The diminutive elf owl, although found elsewhere, is adapted to and dependent on the deserts of the Southwest and the stands of saguaro cacti for its feeding and breeding requirements. Where grasslands, high deserts, and open rangeland prevail, the long-legged burrowing owls keep company with excavating mammals that open up nesting habitat for them.

Assigning other owls to specific habitats is not quite so absolute, but while the ferruginous pygmy owl does favor the openness of the mesquite and live oak forests in its limited range in the southwestern United States, its close relative the northern pygmy owl is partial to the mountainous and open forest habitats of the Far West.

The whiskered owl's population in the American Southwest is closely associated with stands of sycamore found at higher mountain elevations, where it is occasionally joined by flammulated owls, whose range in the West is tied to the open, dry, and mature montane conifer forests.

Spotted Owl (*Strix occidentalis*)

Certainly the spotted owl has been the subject of much controversy—through no fault of its own, but because of the conflict between preserving its old-growth

5.1 Northern spotted owl launching an attack on a bat.

habitat and encouraging timber-industry employment. Except for the orca, no species in my part of the world has received as much press as this bird. The vigorous objections to cutting the last fragments of ancient forest in the Pacific Northwest, with good reasons, cite the owl's dependency on these remaining pristine conditions for its existence.

Among my encounters with wild northern spotted owls, one occurred in the mixed ponderosa pine and Douglas fir stands along the northeastern slopes of the Cascade Range. Forest Service personnel had agreed to take my good friend the artist Thomas Quinn and me into the bird's secretive haunts so we might get a very close look at the owls and share a few moments with them. I recall Tom and I commenting on the circuitous but always intriguing route our guides took to get us in range of the parent owls and their youngsters. For several understandable reasons the trail into the nest site was indirect. Although there was much to see and admire as we hiked in with our two guides, it seemed we took two steps forward and one back as we made our way. This of course made it impossible for us to retrace a route leading to the nest. We would not be back to disturb the birds—this was a one-time visit. Another reason to obscure the nest site was more ominous. Any knowledge of where the owls might be found could allow word to be passed along, and soon the area could be flooded with visitors, to the detriment of the birds. Or worse, not unknown to occur here in Washington, someone would shoot the birds, thereby ensuring that these old-growth trees had no population of owls to stand in the way of cutting them down.

Crossing a clear-cut, we entered an edge of old-growth trees and soon came into the company of an adult owl with a wide-eyed youngster. Close by, we were told, was a cluster of mistletoe, the top of which had served as the owl's nest. The birds showed no particular alarm as we stood silently, some thirty feet away from their perches a dozen feet up into the bare branches of a ponderosa pine. The isolation of this species has rendered them somewhat fearless, and initially the owls seemed as curious about our appearance in their woods as we were in studying them up close. Heads bobbing up, down, and back and forth, they got a good look at us as we snapped pictures, stored away mental images, and sketched them to record as much of their beauty and manner as possible. After a quarter hour, they seemed to have had enough of us and began to retreat into the deeper woods as we made our way back out toward the clearcut slope. Hiking out, I remember being stricken with anger and sadness. Here was a species on a path to extinction, and the forces leading to it are all of our doing. Perhaps I would never again share its company.

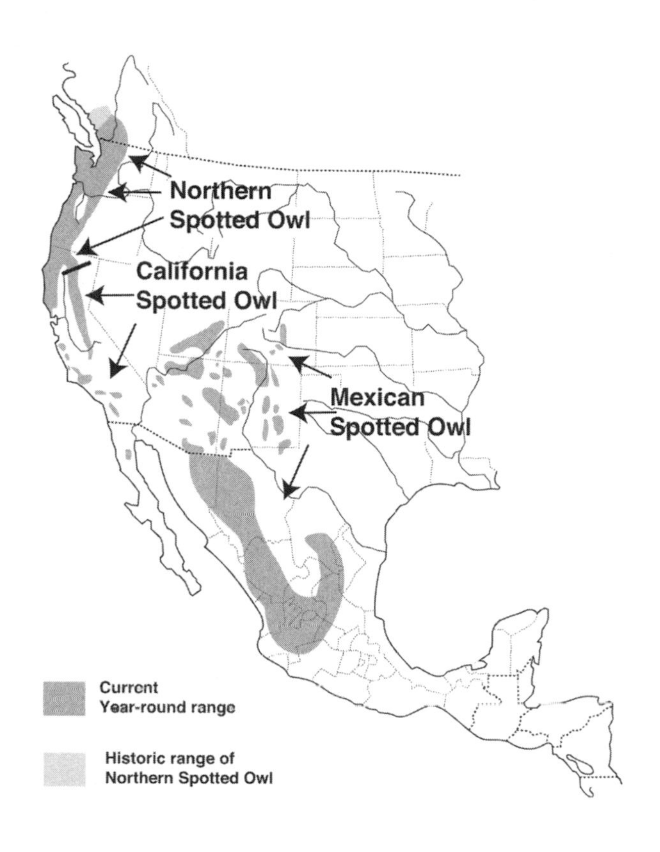

5.2 Range map of spotted owls in North America.

RANGE AND HABITAT

The reclusive northern spotted owl's future is almost exclusively dependent on the ancient conifer forests of the Far West. The other subspecies, the California spotted owl and the Mexican spotted owl, extend the spotted owl's range southward in fragmented patterns from the northern California coast range eastward into wooded canyon country of Utah, Colorado, Arizona, and New Mexico. Farther south the birds are found in the isolated Mexican mountain ranges of Sierra Madre Occidentalis and Sierra Madre Oriental.

Studies of these owls suggest that part of their dependence on old growth is due to the fact that these multistoried forests respond to the changes in ambient temperatures. As temperatures increase, the owls retreat to the cooler portions of the dense canopy to escape the heat that would be more intense in an open and younger forest. The owls move within this complex arboreal environment to find a microclimate where the temperature conditions are most suitable to them. In general, old stands, whether conifers or mixed conifer and deciduous trees, are essential to all subspecies of this owl for roosting, nesting, foraging, and, in times of plenty, food caching.

FOOD PREFERENCES

As in their requirements for nesting and roosting, all subspecies of the spotted owl are dependent on the animals found in the mature forests. Northern spotted owls favor northern flying squirrels and both dusky-footed and bushy-tailed woodrats. They also eat red tree voles, snowshoe hares, and pocket gophers, along with the occasional bat taken opportunistically. Other birds and insects make up only a small portion of their diet.

VOCALIZATIONS

An advertisement and territorial call is a "hoo-hoo-hoo-hooo," intended to establish and define its occupancy in a forest. When courting, the owl gives a softer version of this call, also serving as a contact call between pairs of birds.

A bark and an "ow!- ow!-ow! ow!" serve as a territorial defense call, typically given at sunset but also used throughout the night as needs require.

A nonvocal bill clicking or snapping occurs when these birds are agitated.

COURTSHIP AND NESTING

In the northerly portions of their range these monogamous owls begin their advertising/courtship calls in February and March, and by mid-March or early April a nest site is selected—often a mistletoe broom, tree cavity, old raptor nest, or, in the case of the Mexican subspecies, a pothole or rock ledge.

As in other owl species, the female alone does the incubating of the eggs and brooding of the young birds, all the while dependent on the male for sustenance. Hatching occurs some thirty days after laying, and by the thirty-fourth day in the nest the youngsters are developed enough to fledge. For another two months or more the youngsters remain in the care of the parents, as they continue to feed them at a rate that diminishes as the young owls develop their flight and hunting skills and gain independence.

Spotted owls can be vigorous defenders of their families and show no hesitation to attack goshawks, ravens, pine martens, and even humans that appear to pose a threat to their offspring.

THREATS AND CONSERVATION

Goshawks, great horned owls, barred owls, and pine martens will take the young spotted owls as prey. Also, starvation, particularly over their first year on their own, is a threat to the young owl's welfare. Habitat loss, however, remains the pervasive force that continues to erode and decimate this species' numbers.

In the Pacific Northwest the little pristine habitat remaining to these birds is so fragmented that in some areas the more adaptable barred owl has expanded its populations and displaced the spotted owls. Over the past half century, the barreds have occupied the compromised forests and aggressively preyed on young spotteds and interbred with the adults, threatening to dilute their strains with hybrids that are capable of reproducing.

The U.S. Fish and Wildlife Service has concocted a plan for shooters to kill nearly four thousand barred owls in an effort to limit their impact on the spotted owl. This is more like shooting the messenger than solving the problem. The adaptable barred owls will be back in numbers, and without a viable habitat to live in, the spotted owl will be overwhelmed by its bigger and more aggressive cousin. Historically, forest management has favored economic interests and has only recently focused on restoring and preserving the ecological services provided by ancient forests. Wildfire management has likewise been neglected to the point that, when fires do occur, they burn with a breadth and intensity that is devastating to forest-dwelling species, particularly the spotted owl.

The Forest Service did reduce the logging of old-growth forest dramatically in the 1990s, and this was a positive gesture, but it still may have been too little too late. The fate of the northern spotted owl is worrisome.

VITAL STATISTICS

The northern spotted owl is long lived. In one study, 123 birds were banded, and four males and four females lived at least twelve years. Oregon recorded owls living to be sixteen and seventeen years, with one bird still alive after twenty-five years.

Length: 16.5–19 inches (42–48 centimeters)
Wingspan: 40–50 inches (100–125 centimeters)
Weight: 1.3–1.4 pounds (590–635 grams)

Elf Owl (*Micrathene whitneyi*)

I grew up reading some of the early accounts of nineteenth-century naturalists as they collected specimens for museums and later lugged their bulky box cameras in the first concerted attempts to provide a photographic record of a species. When William Leon Dawson began to write his first books on the birds of particular states, Audubon's monumental work had been, for more than sixty years, the go-to source for images and information on the birds of America. Dawson, lugging his equipment in backpacks and wagons, had completed his bird books for Ohio and Washington when he produced his four-volume set on the birds of California. It was in this work that he described what it took to get his work done with the elf owl in what might be termed today as particularly florid terms. Extravagant or not, his words do capture some of the intrigue and challenge a naturalist might face when seeking to become familiar with a wild creature.

From *The Birds of California:*

> The oologist will pack a ladder for weary miles over the desert. For this he will invade the haunts of the "side-winder" and the Gila monster. For this he will wrestle with tediously unending creosote and insinuating cat's claw. For this he will brave the cruel cholla, which looses its bunched lances at a touch, or pierces the feet of the passerby. For this he will ascend rickety heights of saguaro: if need be, hug its spiny column to meet a flaw of wind or to gain an objective just six inches higher. (The thorns can be removed from the knees and arms at leisure over the campfire.)

There are often curious and fascinating stories associated with the naming of birds, and so it is with the elf owl, whose Latin name, *Micrathene whitneyi,* is given to the bird out of respect for Josiah Dwight Whitney, an esteemed professor, field geologist, and early surveyor of what is today the American Southwest. Whitney didn't discover or describe the species; this was the work of James G. Cooper, whose exploits Whitney supported. He also has the unique honor of having both the highest peak in the contiguous United States, Mount Whitney, and the smallest owl in the world named after him.

RANGE AND HABITAT

About the size of a song sparrow, this tiniest of owls in body mass (the least pygmy owl is smaller in length, but heavier) is a very abundant raptor in the upland deserts of

5.3 Elf owls scolding a ringtail.

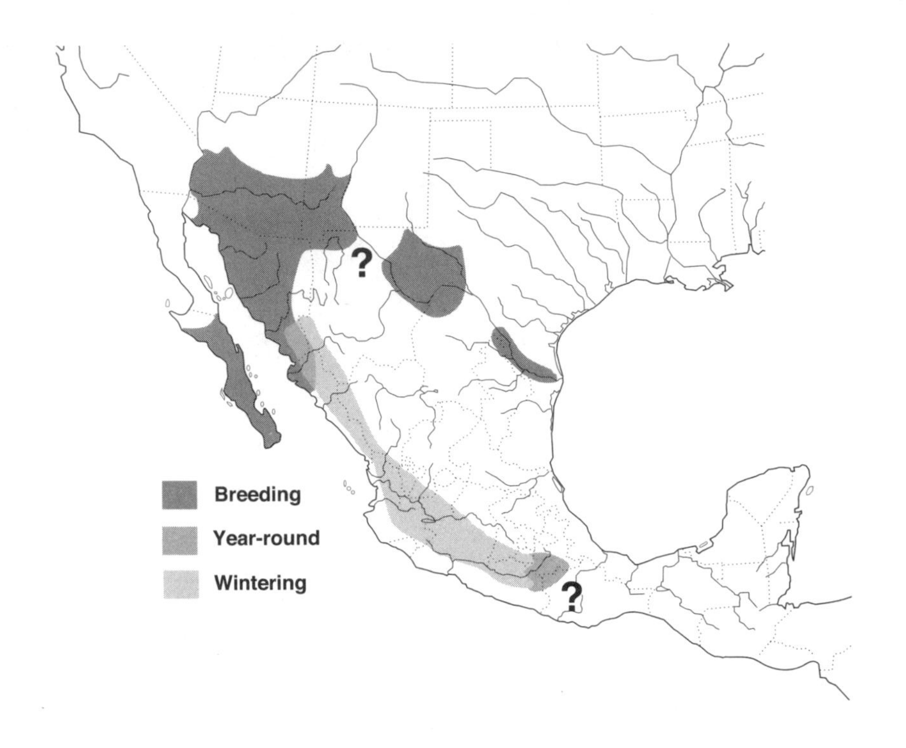

5.4 Range map of elf owl in North America.

Arizona and Sonora, Mexico. Its breeding range includes Arizona, New Mexico, Texas, and Baja and northern Mexico. Wintering populations extend through Mexico to the edges of Central America. Throughout its range the elf owl favors the lowland desert along wooded washes and up into adjacent riparian mountain woodlands, where it will nest at elevations from 3,300 to 6,500 feet (1,000 to 2,000 meters).

In parts of its riparian woodland range, where ample prey and nesting cavities are available, this species shares the habitat in close company with a guild of other small owls, including the whiskered, flammulated, northern pygmy, and western screech owls.

FOOD PREFERENCES

The elf owl especially favors arthropods, including moths, beetles, and beetles. Like the whiskered owl, this species is skilled at removing the stinger from scorpions. Vertebrates also are part of the diet, as the owl will catch blind snakes, lizards, and mice. When it finds an abundance of these latter prey species, the elf owl will often cache them in a cavity for later consumption.

VOCALIZATIONS

Dave Ligon, Susanna Henry, and Fred Gehlbach describe a wide range of sounds these diminutive birds are capable of vocalizing.

A chatter song composed of a yipping, not unlike a small puppy, is part of the elf owl's territorial advertising and courtship singing repertoire. The male often offers it to the female as he perches in the entry to a prospective nesting site he has selected.

"CHUR–ur—ur" is also a distinctive male owl call, as the bird flies from the nest cavity he is advertising. In the process of courtship the female produces a "rrrrrr" trill when the male feeds her and "sheee" vocalization when copulation occurs. Once the youngsters have hatched, they soon peep and squeak and can be heard from outside the nest at some distance.

Both the male and female elf owl employ a barking call directed at intruders to their territory, which sounds like "cheeur—cheeur" and is delivered in rapid succession two to three times. The pair also sustains contact with each other with a soft, whistle-like "peeu" call.

UNIQUE STRATEGIES

Like other smaller owls, the elf owl will assume an upright and sleek posture that blends with its immediate environment. Also, like the whiskered owl, these birds will feign death when handled, an adaptation that encourages a predator to relax its grip or be momentarily distracted, at which point the bird can make its escape.

The northern populations of elf owls vacate their habitat to migrate farther into Mexico. These owls are not as well insulated as the western screech and whiskered owls, so it is in their interest to seek warmer temperatures and more abundant arthropod populations farther south in winter months.

When nesting, these birds are unique among small owls, because they have been observed removing material unsuitable to them left by house sparrows that had previously used a ladder-backed woodpecker cavity before the elf owls took possession. Also, in an Arizona study, these birds were found to have 95 percent hatching efficiency of their eggs, the highest among small owls.

Like the whiskered owl, the elf owl will nest in close company with other small owls, as well as with a rich diversity of other cavity-nesting avian species, including trogons, flycatchers, and tits. This diminutive bird will dive at the head of a perceived predator, and if an entire community of nesting birds spots an intruder, they will descend on it like a swarm of bees.

COURTSHIP AND NESTING

At lower elevations the male owls arrive on their breeding grounds as early as the middle of February, and at higher elevations as late as the middle of April. Advertising of nest sites commences as the females arrive, and when a female accepts a location, the two form a monogamous three-month, one-season pair bond.

Depending on temperatures and elevation, the average clutch of three eggs may be laid from early May into June. As is the custom, the female does all the incubation and brooding of young, with the male providing the sustenance to his mate and hatchlings. After a little over three weeks of incubation, the eggs begin to hatch, and by the fourth week of life or a few days later the young are ready to fledge. Unlike other small owls that have growing to do after fledging, these birds leave the nest at adult size and fly, albeit weakly, out to perches beyond their nest cavity. The recently fledged owls readily catch live crickets.

In their desert habitats where they occupy stands of saguaro cacti, the elf owls are almost entirely dependent on the cavities dug by gilded flickers and gila woodpeckers. More recent research has revealed abundant populations of elf owls nesting in the cottonwood, Arizona sycamore, and white oak mixed forests established along canyon slopes and streamsides. Here too, the owls' dependency on cavity-digging woodpeckers for nesting sites is great, as are the other species of small owls found here.

THREATS AND CONSERVATION

Gopher and rat snakes, along with ringtail cats, will catch young owls. Great horned owls, Cooper's hawks, and gray-breasted jays are the larger predators. A more pervasive natural threat to this species can be exposure to severe weather and a short supply of the arthropod prey base it depends on for food.

When saguaro cacti are dug up, cut, or otherwise removed, the birds' nesting locations are further limited. Likewise, in woodlands of the Southwest, fires can decimate the birds' nesting requirements of stands of riparian woodland.

The human effect is felt when the uninformed or indifferent birding public pounds on nest cacti or trees to flush an owl so they might see it. The disruption can be serious enough to cause the abandonment of the eggs or brood. Playing phone app recordings to determine if owls are present is also a problem. The key, again, is education; anyone who knows the effect such a disturbance can have on the owls is unlikely to commit such an error.

Along with environmental education, posting signs, and distribution of brochures, there are also other proactive projects that may be undertaken. Where burned or damaged forests have reduced breeding opportunities for these birds, nesting boxes may be introduced under the guidance of the local department of wildlife. Short educational programs sponsored by local conservation groups or an enlightened service or business group can also support such projects, as well as volunteers who wish to share their understanding and passion for owls.

VITAL STATISTICS

There is a record of an elf owl living for fourteen years in captivity. A report of a bird living for nearly five years leaves me thinking that more work will be required to get an accurate measure of this species' average longevity.

Length: 5–6 inches (13–14 centimeters)
Wingspan: 13.5–16.5 inches (34–42 centimeters)
Weight: 1.2–2 ounces (34–56 grams)

Burrowing Owl (*Athene cunicularia*)

The cardboard box I was handed sounded as if it contained a demented gnome pounding the interior with a pair of drumsticks. This was the opening moment I well remember when a burrowing owl was delivered to me. It had been rescued from someone who had no idea what was necessary for its care and feeding. A dynamo in feathers, it was well adapted to the rugged and competitive life on the ground of the high desert of eastern Washington, where it likely came from, but wanted no part of any rehabilitation I might be offering. Sharing territory with scorpions, rattlesnakes, badgers, and coyotes, I believe I was considered just another untrustworthy, tough terrestrial being to be feared.

This bird had had its wings clipped, but when I opened the top of the box he bolted out as if spring loaded and sprinted like a stub-tailed roadrunner across my studio. Before I even headed in his direction, the owl had retreated to a dark, narrow space behind my bookcase where, with a flashlight, I could see him pressed into the deepest recess and glaring back at me with defiance. Unlike the more inquisitive and tolerant great gray and spotted owls that I knew, it was clear that gaining even a modest portion of this owl's trust would be a long process.

5.5 Burrowing owl with captured kangaroo rat.

Over the time the burrowing owl lived with me, its shyness never diminished much, and I was left with the feeling that this species, unlike others of its order that I've had in my home, has little capacity for tolerating people. To provide the proper space the burrowing owl needed, I built an enclosure that included an underground retreat along with a post and a few small boulders where it could feed from and sunbathe. Whenever it spotted my husky or me, however, the owl would dash into the entrance of its subterranean refuge and disappear. It moved like a ground squirrel eluding a hunting eagle. From time to time I'd wait him out, and eventually a face would appear at the burrow entry and he'd cautiously emerge.

On a particularly cold and rainy day I went out to feed the owl and, not seeing him, simply left his food on top of the post. Returning several hours later, however, the food was still in place and no sign of the owl. I began to be concerned, because raccoons have occasionally raided my bird enclosures, but I couldn't determine how

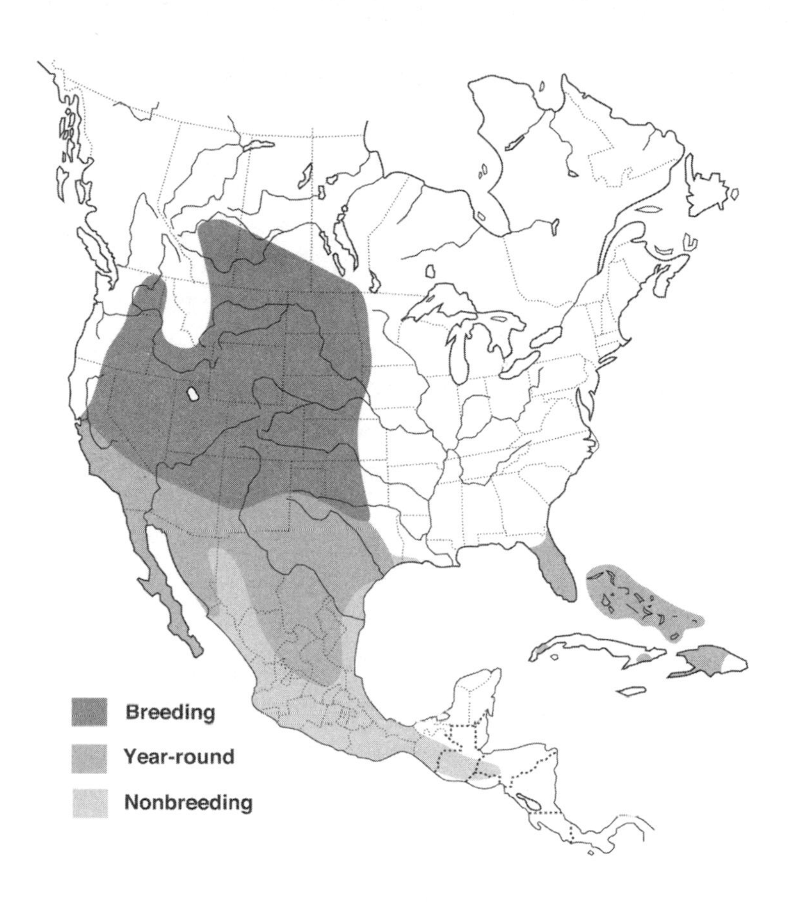

5.6 Range map of burrowing owl in North America.

they might have gained access to get the burrowing owl. Curiosity got the better of me, so I opened the cage and, with my hand, began to explore the passageway of the bird's burrow. No sooner had my extended arm darkened the entry than a very loud and sustained rattling sound, like castanets, came from inside the cavity. It sounded so much like a rattlesnake's warning that I reflexively yanked my hand out with sufficient momentum to send me reeling backward. Whether innate or learned, this imitation of the venomous reptile was sufficient to keep me from getting too close.

RANGE AND HABITAT

The Greek and Latin name for this species is well chosen. *Speotyto* is a combination of Greek terms for cave ("speos") and "tyto," meaning owl. The second term, "cunicularia," is derived from *cunicularius,* meaning burrower or miner. Burrows are indeed essential to the bird's reproductive life where it lives, in western grasslands and portions of the Florida peninsula, with populations stretching down into the Caribbean Islands and farther down the continent into Mexico and Central America.

As the native grasslands have diminished in America, the owls have found suitable habitat in cemeteries, airfields, agricultural lands, and expanses of college campuses and fairgrounds. Sharing company with burrowing mammals also seems to be an essential part of the owl's ability to establish subterranean nesting sites. As prairie dogs, badgers, and gophers do their digging, the birds use portions of such cavities for nesting vestibules.

FOOD PREFERENCES

Watching diurnal hunting burrowing owls was a regular part of boyhood expeditions to the Imperial Valley in California. We would follow them as they hopped, ran, and flew over the open ground where, depending on the time of day, they caught scorpions, beetles, grasshoppers, and assorted reptiles and amphibians, including fence lizards, garter snakes, turtles, frogs, and toads. Among the mammals they consume are wood and kangaroo rats, voles, and mice.

VOCALIZATIONS

Along with a singular ability to produce the sound of a rattlesnake, burrowing owls have an impressive array of other calls. Among them are "coo—cooo," the primary territorial song, given only by the male; a warble, delivered by the female during mating as the male also sings a variation of his "coo-cooo" song at this time; a rattle call, which I knew well, suggestive of a rattlesnake and uttered when threatened; rasp calls, given by young birds and brooding females; an "eep" call given by young if stressed and to beg food two weeks following hatching; clicks, screams, and chatter, given in nest defense; and bill-snapping to register irritation and in face of threat.

COURTSHIP AND NESTING

Burrowing owls are monogamous for the most part through the nesting season, which begins as early as March and extends into May. These enterprising owls use the abandoned excavations of badgers, marmots, and prairie dogs, but they are also able to do digging of their own using their beaks and toes to excavate several feet into the ground in only a few days.

A downward sloping tunnel as long as six feet serves as an entrance. It ends with the entry to an enlarged chamber where the female broods her eggs. The owl sometimes uses cattle dung to line portions of the chamber, which is thought to be a possible strategy to conceal the bird's scent from predators. Around the burrow itself, the cattle manure is thought to attract dung beetles that the owl will then eat.

As many as seven to nine eggs will be incubated only by the female for up to thirty days, and by the end of their second week from hatching the youngsters emerge to take their first look at the world above ground. As in other species of owls, the male is the sole provider of food to the family during incubation and the early brooding of the hatchlings. By their sixth week the owlets are flying about near the nest and are essentially fledged.

THREATS AND CONSERVATION

The digging mammals they share habitat with are also among the principal predators of burrowing owls. Skunks, weasels, opossums, and badgers will predate families of these birds. The local great horned owls, prairie falcons, and Swainson's, red-tailed, ferruginous, and Cooper's hawks capture them as well.

The burrowing owl's numbers are declining, and the widespread threat to this species is expanding agricultural interests. Moving into formerly dry country, agriculture gets a boost from irrigation, turning burrowing owl habitat into vast fields of sunflowers, sugar beets, or potatoes, to mention a few of the crops. The herbicides and pesticides that are applied in crop production likewise are a threat, because these poisons work their way up through the food chain and compromise the health of the birds.

To mitigate the loss of this lovely owl, we must consider not only reserving suitable landscapes for it to live in, but also to see that conditions are appropriate for its prey base and the presence of the digging mammals it coexists with and depends on for breeding locations. Limiting or rerouting traffic that crisscrosses its flyways and hunting territories will certainly reduce the number of fatal collisions of burrowing owls with automobiles.

This species can be introduced successfully into new areas and fortunately does seem to respond to the placement of underground nesting boxes where excavating mammals are absent but suitable prey for the birds is available. Here, too, observation posts can be provided for the owls to perch on in their prey search and scanning for possible predators. No poisonous pesticides should be applied closer than 800 feet (250 meters) of the owl's nesting locations, to help avoid exposure to the birds.

Of course, education can play a singular role in elevating public interest and stewardship of the species. Diurnal in its habits, the bird is easily observed at a respectable distance, providing a thrilling first-hand experience for classes of students and groups of adults. Science concepts related to carrying capacity, interspecific behavior, adaptation, and predator-prey relationships are only a fraction of the subjects that can be explored in such a "live in the field" experience. The aesthetic pleasure of seeing and hearing the birds in their native settings is immeasurable.

VITAL STATISTICS

Not much is known about the longevity of this species, although one banded owl is reported to have lived for at least eight years.

Length: 9–11 inches (19–25 centimeters)
Wingspan: 20–24 inches (50–60 centimeters)
Weight: 5 ounces (142 grams)

Ferruginous Pygmy Owl (*Glaucidium brasilianum*)

For those of us who are fascinated with the avian world it sometimes comes as a surprise to learn that compiling detailed information about a species requires that some birds be sacrificed. Size and weight measurements and details of diet and overall health can't be determined by observation alone. Many of our earliest artist-naturalists also benefited from having the bird in hand, whether it was alive or dead. George Sutton, like the great bird illustrator Louis Agassiz Fuertes with whom he studied, was able to draw and paint first-hand impressions of his subjects as he collected them on the numerous scientific expeditions he took in the United States and Mexico. Many of these artistic impressions have a vitality otherwise lacking in much of contemporary bird art. Simple as they may be, they are informed by Sutton's field experience with his subject and the excitement and respect he felt for the bird in hand. It is not stretching the truth to say I can see in some of his work his emotional connection to the subject and the location in which he found his specimens.

An episode of collecting a ferruginous pygmy owl described in Sutton's book, *Mexican Birds: First Impressions,* more than hints at the investment the artist may sometimes make in developing his statement on the subject. Exploring the forests of the Yucatan Peninsula, he comments on the oppressively hot weather and that "The

5.7 Ferruginous pygmy owl pursuing a frog.

ticks were getting us down." And although the excitement of collecting rare specimens helped buoy their spirits, "The daily grind of picking the pinolillos off, eating sulphur, shaking and hanging out clothes, and applying iodine to the bites was nerve-racking and fatiguing."

Returning to his camp at sunrise from a night of searching for nighthawks and whippoorwills, Sutton caught sight of a hummingbird diving repeatedly over a low-growing palmetto. To his astonishment, as he watched, a "small brown knob" appeared to emerge from the leaf stems, then shrink back each time the hummingbird attacked in its direction. With binoculars Sutton inspected the knob and discovered it was a "hunched up, golden-eyed ferruginous pygmy owl." Sutton's further description of the moment, from *Mexican Birds: First Impressions,* is worth sharing:

> While watching developments, I noted that the owl's eyes were piercingly bright, and that the hummingbird was beside itself with animosity. The owl was not tucked in for the day, as orthodox owls should be by sunup, but was out for a hunt and hungry as a tiny bear. Why it did not try to catch the hummer I do not know, possibly it was too intelligent for such a waste of energy. Its incredible smallness made the hummer seem unnaturally large and fierce. When, with roaring wings and flashing gorget, this thimbleful of fury bore down in attack, the owl cringed visibly.

Later in the day Sutton came upon another ferruginous pygmy and, needing a specimen, sought to collect it by shooting it from its perch over a small river. The scientist watched with dismay as the bird fell into deep water some distance from shore and began drifting with the current. Taking off his clothes in preparation to swim for his prize, Sutton was further alarmed when he saw a fish surface and eventually pull the owl beneath the water. Fortunately, the owl bobbed out of the fish's grip, and in this moment Sutton felt his only recourse was to shoot his aquatic competitor, which he did before plunging into the river to retrieve the bird.

Recovering the owl from the stream was only the beginning of the challenges he faced in saving this valuable specimen. A saturated bird, of course, doesn't dry quickly or recover its feathered contours without constant fluffing and shaking. This was a dead bird, so Sutton set to work putting the feathers back into place on the body so that it did not appear too badly damaged by the fish bites. It was not an easy task in the hot and humid climate that was also hastening the decomposition of the owl's body and internal parts, making skinning and examination of the owl's diet all but impossible. Suffice it to say, however, he succeeded.

RANGE AND HABITAT

In the United States these owls have a presence only in southern Arizona and the most southerly tip of Texas, because they are mainly dispersed into Mexico and

5.8 Range map of ferruginous pygmy owl in North America.

Central and South America. The southeastern Arizona population is normally found in the riparian woodlands and mesquite habitats between 1,000 and 4,200 feet (300 to 1,300 meters) elevation. In the most southerly edges of Texas, this pygmy owl does well in the live oak and mesquite forest of the Wild Horse Desert. Its populations in Mexico and Central and South America are established along the edges of a variety of tropical ecosystems below 4,000 feet (1,200 meters).

FOOD PREFERENCES

Insects such as grasshoppers and crickets are its preferred diet, but in Texas and Arizona these owls also catch reptiles, birds, and small mammals. Although they are best described as perch-and-pounce hunters, the owls sometimes investigate the nesting cavities of smaller birds to take the young; should a cavity be vacant, it will be used as a cache site for the prey they have caught.

Like its closest relative, the northern pygmy owl, these birds are capable of taking prey as large as themselves or larger and have been noted taking a Gambel's quail and cotton rats.

VOCALIZATIONS

A "pee-weet" call is used as an alarm if the owl is threatened, or if its territories are intruded upon.

Chittering calls are used by the young as they beg to be fed. The female owl emits this same call when soliciting a meal from her mate.

"Toot" calls are produced by the male to advertise his territory to females and are given continuously with five- to ten-second pauses between them. The persistent male may continue to use this call for up to five hours in the process of establishing his territory and courting the female.

A "chu, chu, chu," offered up from a favored perch, is also used in the owl's advertisement of territory.

COURTSHIP AND NESTING

In Texas this nonmigratory owl will maintain its territory throughout the year. The area around the nest site that it defends may cover a radius of 1,100 to 2,000 feet (350 to 600 meters). The defense of territory begins in earnest as early as January, as the male introduces the female to possible nesting sites. These birds prefer the cavities of woodpeckers in stands of oak, cottonwood, mesquite, ash, and saguaro cactus.

By March the female has made her choice of a cavity and will lay, on average, three to five eggs and begin incubation immediately after the first egg is laid. By mid-April, over a span of twenty-five to twenty-seven days, the eggs will have hatched. The female alone incubates and continues to be the sole brooder of the young owls for another three weeks. Throughout the time the female has the nest responsibilities, the male is the sole provider of food.

By their fourth week from hatching the youngsters are fledging. In another eight weeks, when the young owls have developed their foraging skills and no longer depend on their parents for food support, they begin to disperse from the parental territory.

THREATS AND CONSERVATION

The ferruginous pygmy owl faces some of the same predators that other small owls do, particularly Cooper's hawks and great horned owls. Raccoons will raid their nest sites for eggs and young.

A more extensive threat, arguably not a natural one, is the gradual and continuous fragmentation of their habitat by human enterprise. Particularly in the American Southwest, feeding and breeding options are reduced as housing and recreational development along with agricultural expansion impinge on their habitats. Conversely, where the bird's environment remains intact, these owls appear tolerant to a modest degree of human disruption. When people have inspected nests, the females readily return to their incubation duties.

It remains unclear to what degree the ferruginous pygmy owl might benefit from nesting boxes placed in areas where their options for natural cavities have been reduced. There is, of course, no question that, where human interests and the owl's activities meet, educational information on the species can be of value. Although not drawing direct attention to locations of nests or hunting haunts, discreet references can be made to the importance of sustaining the owl's environmental requirements and the proper protocols of behavior when in the bird's presence.

There are four widely ranging subspecies of these owls, which suggests that some of them are more adapted to surviving in specific habitats than other more isolated groups. Understanding these adaptations may help in conserving and sustaining habitats and general living conditions for the species.

VITAL STATISTICS

There is not a lot of information on the longevity of these birds, but studies of nesting pairs of banded owls in Texas indicate that they nested in the area where banded for four years.

Length: 6.5–7 inches (16–18 centimeters)
Wingspan: 14.5–16 inches (37–41 centimeters)
Weight: 2–3 ounces (56–85 grams)

Northern Pygmy Owl (*Glaucidium gnoma*)

In our house one winter, soon after the lights were out, we'd begin to hear a buzz-like whirr of wings moving from room to room. If it entered our bedroom, as it did

5.9 Northern pygmy owl with captured California quail.

occasionally, I could feel the slight turbulence created overhead by the passing of a miniature form. Circling briefly, it would gain momentum and at greater speed plunge itself into the darkness of our hallway to continue its explorations. Such was our nocturnal life during the short period of time a northern pygmy owl lived with us.

Captured by a licensed bird bander in a live trap set for falcons, the little owl was in need of some rehabilitation and given to me for a short period, allowing me to both study it and prepare it for release back to the wild. The diminutive but fierce pygmy owl would accept no containment within an enclosure, and it quickly became clear that the limited freedom of our entire house would have to be its only confinement. To deal with its roving inclinations we placed newspapers beneath curtains, the tops of which became favorite perches, and over the backs of furniture. A wet towel and a mop were constantly on hand.

At this time, social occasions at the Angells' were a little out of the ordinary and often included fly-throughs by kestrels, merlins, and owls. In fact, my good friend the artist Fen Lansdowne would often venture from his home in Victoria, British Columbia, to join me in what was essentially a living-room aviary where we could watch these species in close-up action and inaction during our conversations.

My home being large enough, I occasionally introduced prey for some of the birds I rehabilitated, knowing that if they were to get on with their lives their predatory skills would need both development, for the younger birds, and honing, for the adults in my temporary care. This northern pygmy owl in particular showed no hesitation when hungry and would launch from the top of a curtain and snatch a scrambling white mouse, dispatching it and pushing off to find an aboveground perch where it could feed. This activity required some adjustment, and unless the feeding perch was properly buffered with towels and newspapers, the process could get pretty messy.

The owl was soon in shape to be released, and we decided to do so here at my home, thinking that the setting was natural enough that the bird would transition back into its normal habitat without a problem. Should the owl find difficulty, I felt that keeping an eye on it would allow recapture and further rehabilitation. I need not have worried. Within an hour of its release, feathers were floating out from the middle canopy of a cedar tree along the trail to our creek, and looking up we spied the little owl plucking a freshly caught junco.

RANGE AND HABITAT

Northern pygmy owls are found from southeastern Alaska southward into the Rocky Mountains of British Columbia and Alberta in Canada and into the interior and coastal ranges of Oregon, Washington, and northern California. Their populations are established as well in the more mountainous regions of Idaho, Montana, Colorado, Arizona, and New Mexico and continue down into Mexico and Central America. This species prefers various forested habitats both in bottomland and higher elevations ranging to 4,000 feet (1,200 meters). As with many of the small woodland owls, trees with cavities, both natural and woodpecker excavated, are essential.

FOOD PREFERENCES

Although other owls similar in size (such as the flammulated and saw-whet) are partial to insects, this avian dynamo's diet favors mammals and birds. They capture a variety of species ranging in size from shrews to red squirrels. Mice, voles, and chipmunks are also included in their diets. This is a determined hunter, because it captures hummingbirds, and one has been recorded capturing a California quail, a

5.10 Range map of northern pygmy owl in North America.

bird more than twice its weight. Their appetites are as big as the prey they catch, because even a relatively inactive nonbreeding captive owl will consume as much as a quarter of its weight a day. The wild Eurasian pygmy owl will consume up to half its weight in prey each day.

Pygmy owls have symmetrical openings in their skulls for receiving auditory cues. Because they are among the more diurnal owls, they are less dependent on hearing their prey than on spotting it and flying it down. Their plumage is likewise more rigid than owls with softer feathers that dampen any sound they might make. The northern pygmy is more akin to a shrike, as stealth has little to do with capturing its meal.

VOCALIZATIONS

The "toot" territorial and nest advertising call of the male is heard early in the breeding season and continued throughout the balance of this period.

A "cheep" vocalization is associated with begging for food by the young.

Alternating toot calls are given by dueting adults.

A short trill and toot calls are uttered when the birds are copulating.

The chatter call is associated with the owl in an agitated state.

COURTSHIP AND NESTING

Although the northern pygmy will move to lower elevations in the winter months, it is a nonmigratory owl that remains with its mate through the breeding season. In late winter the male owl may be courting a mate with song and invitations to inspect the possible nesting cavities he has selected. During the site approval period, the birds engage in dueting, and with a location secured, copulation soon follows.

The female lays from five to seven eggs from April to June, usually in an abandoned woodpecker cavity excavated in an old snag that is part of an unlogged mature forest, which provides a heavy canopy over and around the nest site. The incubation period for each egg is about twenty-eight days, with the male owl providing all the food for the brooding female and (for at least another week after they hatch) the young as well.

By the middle of the fourth week after hatching, the young begin to fledge at intervals of one to two days of each other. Not yet capable of flight, they nevertheless launch from the cavity, seeking some immediate perch or going all the way to the ground, where they immediately seek some point to climb to. Within a few days, however, between hops and flutters, they become capable of short flights and are responding to the parent bird's prey deliveries with begging cheeps.

THREATS AND CONSERVATION

The northern pygmy owl has few predators apart from the tree-climbing mammals it may share portions of the forest with—pine martens and occasionally raccoons. Spotted owls may be one of the few avian predators that consider the pygmy worth making the effort to catch and eat.

Certainly the continued cutting of mature forests of the West will make an impact on this forest-dependent species. The silvicultural practice of removing old snags, which are ideal nest sites and observation locations for hunting owls, is detrimental to their welfare. The effects of blanket spraying of insecticides to control insects in the pygmy owl forests have yet to be fully studied. There is no doubt, however, that these toxins can build up in the systems of some forest vertebrates and, as has been demonstrated unequivocally with the application of DDT, can have a calamitous effect on a species' welfare.

With the construction of expansive mountain houses and their floor-to-ceiling windows for views, owls and other forest birds are paying a deadly price when they fly into these transparent barriers.

Only now are we exploring the possibility of nesting boxes for these cavity-preferring birds. Placing them in logged areas adjacent to stands of trees and providing perching locations in the open areas could suit the needs of the owl perfectly, because small mammals and birds will temporarily occupy the open portions of the forest as new trees grow.

This bird's accessibility as a diurnal hunter makes it a perfect fit for educational programs that take respectful students to the field to observe nature. Although their nesting sites are best kept private, locations where the northern pygmy owls are known to hunt in fall and winter allow for a memorable and motivating visit when sighting them through a spotting scope. Like all owls, once seen they are greatly cherished as part of our natural heritage and never forgotten.

VITAL STATISTICS

A fast and furious life for the northern pygmy owl is likely, similar to the ferruginous pygmy owl, which has a documented life span of four years.

Length: 7–7.5 inches (16–18 centimeters)
Wingspan: 14.5–16 inches (37–41 centimeters)
Weight: 2–2.5 ounces (56–60 grams)

Whiskered Screech Owl (*Megascops trichopsis*)

There is a lot of history to feel as one wanders through the stands of white oak and Arizona sycamore that edge up along the flanks of the Chiricahua and Huacca Mountain canyons. Near the end of the nineteenth century, the Apache people

5.11 Whiskered owl disarming a scorpion.

maintained secret redoubts here as they sought to sustain their sovereignty. The natural history too is unique, for in this region a diversity of bird life can be encountered that is like no other in the contiguous United States. Owls in particular have a presence, and the spotted owl or the great horned owl's basso profundo often proclaims twilight. Through the course of the night, this vocal prelude is followed by a contribution of hoots, whistles, and trills that the smaller owls offer all around. With thanks to the accommodating trees, the cavity-digging woodpeckers, and a varied prey base, elf, western screech, and, at higher elevations, whiskered, pygmy, and flammulated owls are part of this greater ecosystem community.

The whiskered screech owl is among the most recently studied of all of North American owls, and not until early in the twentieth century did much of this species' habits become known to science. Smaller by at least a third than its cousins the western and eastern screech owls, it is also distinct in the deeper yellow of its iris and the many long bristles that extend from the outer edges of its facial disk.

5.12 Range map of whiskered owl in North America.

RANGE AND HABITAT

This small owl is adapted to higher elevations in the riparian canyon forests and can be found at elevations ranging from 3,600 to 10,400 feet (1,000 to 2,900 meters). Early in their identification as a species these owls were considered rare, but in fact they are, in their optimum habitat, among the most abundant of the small cavity-nesting owls.

Nonmigratory, they are established in southeastern Arizona and a small corner of southwestern New Mexico. Their range extends southward into the mountains of Mexico and into portions of Central America. Although they remain on their territory through the year, they may descend to lower elevations during severe winters. These birds require deciduous trees that provide suitable cavities (either woodpecker-excavated or natural), so they will often defend several nest sites from competition for use.

FOOD PREFERENCES

They are decidedly smaller footed than their larger cousins in the genus *Megascops,* and consequently are well suited to catching insects. Arthropods are consumed in great quantities, including moths, crickets, grasshoppers, beetles, and caterpillars. Favoring scorpions, this owl is particularly adept at removing their stingers before making a meal of them. Small vertebrates make up the larger portion of its prey biomass, however, and it catches, along with spiny lizards and blind snakes, brush mice, bats, and shrews.

VOCALIZATIONS

The voices of the various screech owls in North America are distinct features of the species, but, like other members of their tribe, whiskered owls will employ a bark call when an intruder is identified or threatening. Sometimes a hoot call is given and appears to acknowledge the presence of a nonthreatening subject.

A whistle or "kew" or "whew" calls are given between pairs to stay in touch as they move about their territories, and they produce variations on their trills as they copulate and feed their young. Early in the year, the male offers this same vocalization as he advertises his territory or nest site to a possible mate.

UNIQUE STRATEGIES

When hunting in the thick foliage of the tree canopy, these owls will forgo flight in favor of hopping and climbing through the cover. When roosting on a tree limb, they lower their ear tufts, giving a smoother appearance to their body contours, and hunch over the branch's surface to better resemble a knot or a bundle of leaves.

They are among the owls that will feign death when handled, but if the offender's grip is unrelenting, they will readily defecate in an effort to seek their release.

COURTSHIP AND NESTING

In January, selecting suitable nesting locations, the male whiskered owl gives successive trill calls to advertise his territory to a prospective mate. A typical nesting site is a cavity of a northern flicker in an Arizona sycamore; as with other screech owls, the tree itself tends to be set apart from a dense surrounding forest with the ground cover about it being sparse and short. Such a selection allows the owls to make direct flights into the nest entry as well as have a clear view of approaching intruders.

By the middle of March, the whiskered owls are courting and may continue to do so for another month. Typically, the two to four eggs are being laid and incubated by early April into May, and by the middle to the end of May to the middle of June nestlings are apparent. Fred and Nancy Gehlbach, who have studied this species extensively, have concluded that these birds, like the eastern screech owl, may place live blind snakes in company with the owlets to effectively tidy up the interior of the cavity. The snakes eat the fly larvae emerging from the decomposing carcasses of cached prey. Arboreal ants may also occupy the nesting cavity to the benefit of the owls. They consume the carrion while ignoring the young owls, but will readily spray a nest intruder.

The young owls begin fledging in early June and will do so into the middle of July. For at least another month, the youngsters will remain with the parent birds, continuing to be fed as they develop their own foraging skills.

THREATS AND CONSERVATION

Opportunistic hawks and larger owls will occasionally take young whiskered owls, and the eggs in the cavities are sometimes stolen. As in all species, their numbers are reduced in periods of severe weather and starvation resulting from a crash in their usual food supply.

In matters of human-related activity affecting the welfare of these birds, the Gehlbachs' research has given them a particularly good position to assess what measures require attention. These considerations have important implications for other species of owls as well.

Controlled burning in mountain habitat will enhance the likelihood that forest fires will not devastate the riparian forests that this species and other small cavity-nesting species need for reproduction. When clear-cuts (natural or human-caused) do occur, reforestation should be in tree species that the cavity-nesting birds require, such as white oak and Arizona sycamore.

Proper decorum by birdwatchers coming to these regions is essential to the whiskered owl's breeding success. The bird's 85 percent efficiency in hatching its clutch of eggs drops precipitously to 62 percent when the nest is disturbed by people tapping on the tree or climbing it to get a better look at the incubating bird. The young from disturbed nests fledge at a lower body weight, a distinct disadvantage to their survival through their first year.

Limiting access to particular breeding sites will enhance the birds' success at fledging a family. Even a limited closing of roads to these areas will reduce the intrusive effect of automobile traffic on the owls.

As in all programs that seek to enhance the well-being of a species, education of both young and adult visitors to the bird's habitat will have a significant effect in the immediate and long term. Simple signage in the general region of the species briefly describing its behavior, the importance of the forest cover to its welfare, its food requirements, and the proper ground rules for how to behave in its presence will ultimately cultivate a culture of respect and stewardship for the owls.

VITAL STATISTICS

This small owl lives at least four years, but its close relationship to others of the *Megascops* clan suggests that it may live to be much older, perhaps as long as twenty years.

The difference in size between males and females is quite pronounced, with females being 8 percent longer and 14 percent heavier than the males.

Length: 6.5–8 inches (16 to 20 centimeters)
Wingspan: 16–20 inches (30–50 centimeters)
Weight: 2.9–4 ounces (85–113 grams)

Flammulated Owl (*Psiloscops flammeolus*)

It's enough just to stroll into an early summer night in the Cascades to savor the redolence of ponderosa pine, but if patience prevails and moonlight favors you, the initial silence begins to be filled with night birds as silhouettes of the forest are resolved and defined. Such was an evening that I walked along a narrow dirt logging road that penetrated the higher forest stands above the Methow Valley. From somewhere in the shadows an owl called with considerable gusto, and I headed in that direction using the course of the moonlit road as my path.

To one side was a shallow creek with a few trees growing up from its bank and reaching the level of where I walked. This corridor of branches was also the path that a flammulated owl was taking to hawk moths that were about in the moonlight. For perhaps a quarter hour I edged along one side of the owl as it would dart out over the road to snatch at whatever insect might be spooked into flight. For a moment I thought that, like my chickens that often follow me about our yard to see what I

5.13 Flammulated owl with a moth.

might reveal as I garden, the owl was leveraging my disturbance of insects or even small mammals as I walked.

The owl remained only a dark but clear outline, but its long wings and flight manner clearly made it a flammulated. I would not be treated to a call while I watched it, but later, after it flew from roadside to be swallowed up in darkness, I heard a distant hooting that was met with a reply from where the owl I was watching had flown. I wondered if it was the owl's mate, impatient to be fed, as she brooded eggs this time of year.

On a few occasions I've held this species in my hand or it has perched on my finger. They were all undergoing treatment for injuries at a wildlife rehabilitation center outside Santa Fe, New Mexico. Unlike some injured owls I've worked with, these seemed very ill at ease and stressed in spite of the kindness of the handlers. People are not an unknown presence in the lives of many owls, but this highly nocturnal and reclusive species is generally remote from our activities and has never selected for compatibility with us. Like another dark-eyed owl species, the spotted owl, I fear for its future existence as we continue to compromise its habitat.

RANGE AND HABITAT

Secretive by nature, flammulated owls were once considered rare, but in fact they are among the most common owls in their range in western North America. Smallest of their genus, and cryptic in plumage color and pattern, they easily escape notice in the mid-elevation mountain forests from British Columbia southward through the eastern slopes of the Cascades with scattered populations into Idaho, Nevada, and Colorado. They occupy forests through Oregon and down into the Sierra Nevada range in California, with an appreciable density found in the woodlands of eastern Arizona and the edges of western New Mexico. Although records are sparse, the flammulated owl appears to be established through the mountainous pine forests of Mexico.

When the flammulated owl breeds, it has a clear preference for the open, dry, mature montane conifer forests, particularly ponderosa pine. These semiarid forest conditions are also supportive of the arthropod populations that these insect-loving owls require for food. There are also cavity-excavating woodpeckers that provide suitable nesting sites. Some density of foliage provided by trees such as aspen and oak are also important as concealing roosting sites for these secretive birds.

FOOD PREFERENCES

Of all the North American owls, the flammulated may be described as the most insectivorous. Its spring diet may consist largely of moths, and as the days warm, grasshoppers, crickets, and beetles make up the rest. Convincing evidence that these owls feed on vertebrates to any degree is scant. A notable exception occurred in Kelowna, British Columbia, when scientist Doug Cannings discovered shrew remains in the stomach of a collision-killed owl one November. This discovery suggests that the diminutive, small-footed owls may be capable of catching bigger prey when colder conditions reduce or eliminate the availability of insects.

VOCALIZATIONS

Although the flammulated owl does not have a syrinx as complex as that of a songbird, its very flexible labia, combined with its relatively large trachea, permits the production of a remarkable call, despite its diminutive size. The frequency of this bird's vocalization is appreciably lower than that of others of its genus, including the larger western and eastern screech owls and the whiskered owl.

The bird's song appears to be ventriloquial in nature, rendering it very difficult to find in the field and certainly providing an advantage in avoiding predators. It will vary its call depending on the intended recipient, and its prodigious advertising "hoot" carries effectively through thick forest habitat and can be heard at a distance of more than half a mile (one kilometer).

Their secretive habits include retreating to secluded cover where pairs will sing and converse among themselves—they may be audible, but they are difficult to see. In contrast, the owls will come into the open to protest the presence of an intruder near their young with shrieks, barks, and screams. Like all owls, they produce bill snaps when they are threatened.

UNIQUE STRATEGIES

These long-winged owls are capable of flying considerable distances and do so in moving out of the colder higher elevations and northern reaches of their range. By October these birds move south under the protective cover of darkness, seeking more comfortable temperatures and suitable populations of arthropods to eat. Not much is known about where they winter, but by the following April pairs have returned to their breeding territories.

COURTSHIP AND NESTING

With their penetrating voices, the male owls are advertising territories and nest sites for the female's approval soon after their return in April. Although remating is the rule, the birds are monogamous through the breeding season, and courtship involves the male bringing food to the female and the pairs investigating suitable nesting sites from which the female will ultimately make a selection. Preliminary to the female retreating to the nesting cavity to begin egg laying, there is much mutual preening and close association by the owls as they strengthen their pair bond.

Depending on the location within their range, the female may begin incubation from mid-April all the way into late May and will do so in the preferred cavities of northern flickers and pileated woodpeckers. So choice are these sites that, despite the flammulated's diminutive size, they will evict flickers or even saw-whet owls to secure them.

Tended only by the female, a clutch of three eggs will require between twenty-one and twenty-four days of incubation before hatching. Within twenty to twenty-three days from hatching, the young have fledged, clambering about the branches in the vicinity of the nest for several days before they have any capability of flight. As the summer progresses, and in a little over a month after fledging, the young owls have separated from their siblings and parents and are on their own.

THREATS AND CONSERVATION

Occupying the same forest as larger owls and hawks puts a special emphasis on this bird's skills at ventriloquism and secretive manner. Still, great horned and occasionally spotted owls capture them. One has to also consider the predatory potential of the barred owl as it expands its range into flammulated country. This aggressive bird has already demonstrated its effect on the populations of western screech and spotted owls. Goshawks, Cooper's hawks, and pine martens are also natural predatory threats to flammulated populations.

Given its almost exclusive dependence on insects, seasonal variations, particularly cold periods, will limit the owl's access to suitable food.

As far as human impacts, it is important to consider the species' particular function, in an evolutionary sense, which requires the use of cavities provided by woodpeckers in older-growth trees. The cutting of these forests without consideration of the needs of cavity nesters will surely place some portions of their populations in

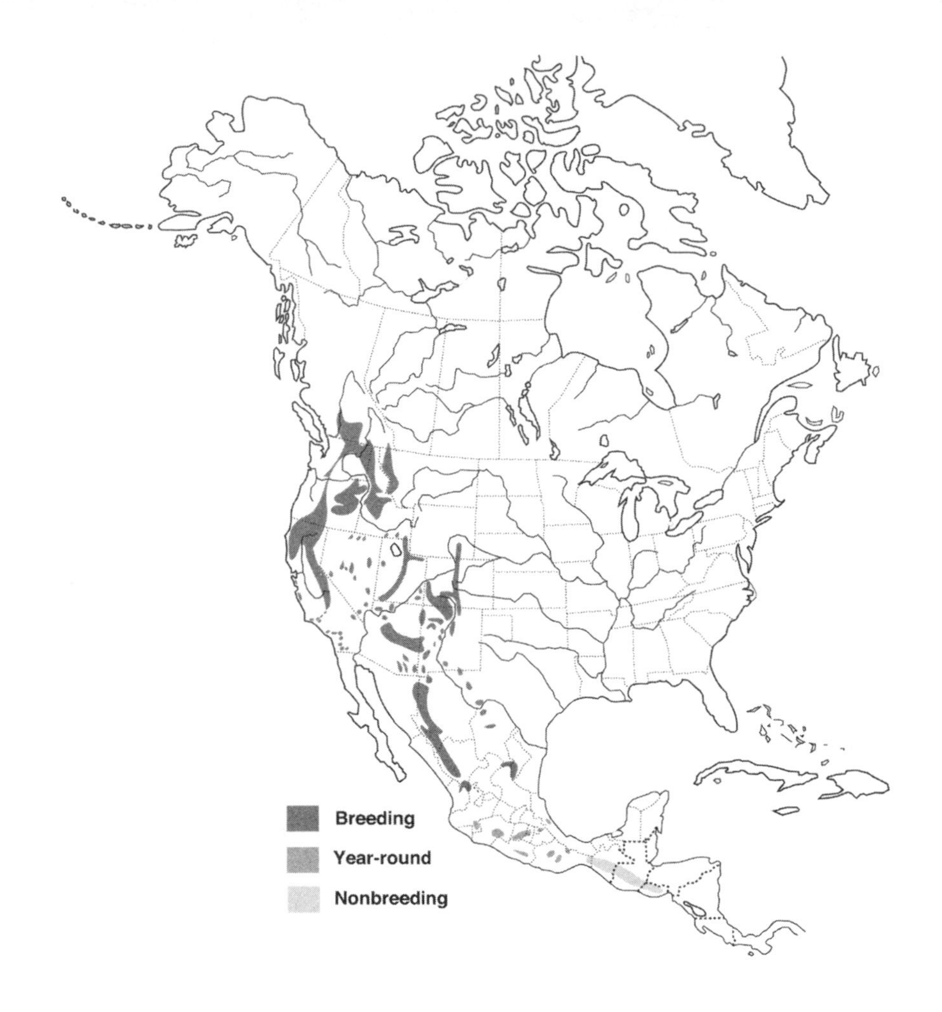

5.14 Range map of flammulated owl in North America.

jeopardy. Furthermore, as owls that have selected for longevity rather than fecundity, they do not reproduce in large numbers, and the spraying in these forests for insect control could compromise the bird's vitality and survival rate by chemical contamination. Pesticides not tailored to a specific pest can also destroy nontarget species that may be an important part of the owl's diet.

Conservation measures are available, and they include the strategic placement of nesting boxes where suitable nesting snags have been removed for firewood or as part of a general clear-cut. The controlled burning and thinning of forests is a very effective forest management technique and assists in returning the forest to more open conditions that are attractive to these owls.

VITAL STATISTICS

As a member of the genus *Psiloscops,* this dark-eyed owl is relatively long lived, with records of males living for more than eight years and females more than seven.

Length: 6–7 inches (15–17 centimeters)
Wingspan: 14–19 inches (36–48 centimeters)
Weight: 1.5–2.2 ounces (42–63 grams)

SIX

Owls of Wild and Remote Places

Only a few species of owls are truly exclusive denizens of wild and remote places—locations where humans have not yet found reasons or capacity to occupy to any extent. Among the most reclusive owls, the northern hawk owl can normally be found only in the far northern forests of North America, Europe, and portions of Asia. Here amid the taiga forests they often share company with boreal owls and great gray owls, the populations of which are, for the most part, restricted to the same regions. Only the snowy owls are the exclusive occupants of the treeless tundra of the arctic, but all of these northern species are extraordinarily well suited by habit and structure to reside in these distant locations so challenging or inhospitable to human encroachment.

Boreal Owl (*Aegolius funereus*)

On my occasional journeys to the foundry that casts my bronze work in northeastern Oregon, I can see the edges of the Blue Mountains just before I drop down into the tiny crossroads village of Umapine. With each visit, the glimpse of those mountains reminds me of my first meeting with a boreal owl.

My family and I had started one of our cross-country trips and, road weary, stopped at a pull-off that took us some distance from the highway. I walked back into the stand of trees and was soon so enveloped by the silence and fragrance of the forest that any sense of a highway nearby was lost. While wandering about, a particular snag caught my eye, because an entry to a woodpecker's cavity appeared to have a

6.1 Boreal owl confronting a pine marten.

bird perched in it. It took a moment to resolve the form, but there staring down at me was the unmistakable visage of a big-eyed, broad-faced boreal owl. Our encounter was brief, because within moments the owl retreated to the interior of the cavity and was gone. The meeting was metaphorical, however. The bird has come to represent the peace that wilderness provides. Here I was, urban man, stepping off a corner of my highway and glimpsing the thin edge of what remained of its domain.

When in the foundry, I seek relief by thinking of the tranquility I discovered in the Blue Mountains and my meeting with the boreal owl—such a stark contrast to the Dante's inferno of the metal-casting environment! Here one is amid the ear-splitting clamor of heavy hammers pounding metal, the shower of hot metallic shards from grinders, and the searing heat from the furnace stirring the gaseous air about the melting bronze.

RANGE AND HABITAT

Aptly named, this small owl thrives in the forests of the arctic and far north across North America and into Eurasia. In this country, population pockets extend southward into portions of the subalpine forests in the Rocky Mountains and the Cascade ranges. Mature and older mixed forests of spruce, fir, and aspen with their closed canopies are preferred for foraging, nesting, and roosting.

FOOD PREFERENCES

Feeding mainly on red-backed voles and heather voles along with bushy-tailed woodrats, deer mice, jumping mice, and shrews, they occasionally take a pocket gopher, chipmunk, and northern flying squirrel. They even catch mammals as large and challenging as weasels and snowshoe hares. The birds taken as food in these northern forests include thrushes, red crossbills, dark-eyed juncos, robins, and chickadees. From year to year as prey availability changes, these owls take a modest number of insects as well, particularly crickets.

In both the breeding season and the wintering period, food is cached for later retrieval and consumption. These birds have been observed to be on high alert in the presence of inquisitive gray jays to keep these corvids from raiding their caches.

Exceptionally keen of hearing, these owls, like the larger great gray and smaller saw-whet owls, have asymmetrical openings of their auditory canals on their skulls. Such an arrangement permits them to precisely locate their prey in both the vertical and horizontal planes without actually seeing it. Even beneath the encrusted snow

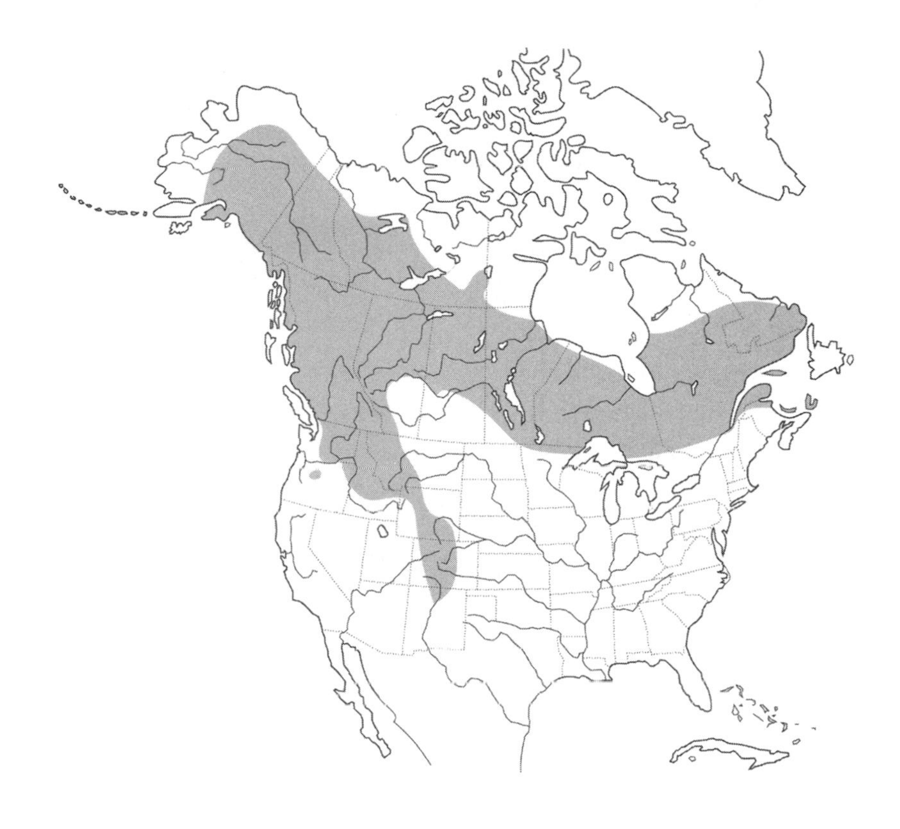

6.2 Range map of boreal owl in North America.

in the older forests they favor, they have access to prey. With their acute hearing they can exactly detect the presence of a small mammal beneath the snow cover and plunge feet first through several inches of snow into the mammal's concealed passageway for a capture.

VOCALIZATIONS

While producing as many as eight distinct vocalizations, the boreal is not easily observed in the act of calling, so the defining of the particular context for some of the calls seems open for further refinement. Nevertheless some of these utterances appear to have a clear function.

By their first week, young boreals are producing a harsh chirp-peeping call that becomes more refined as they age. In the nest the young chatter among themselves as prey is delivered or they aggressively interact. The young owls will be bill-clapping shortly after their first week from hatching, and by two weeks they are capable of hissing.

Although as many as eight vocalizations have been described for the adult owls, their principal songs include the following.

A primary song consists of a pronounced trilling of up to sixteen notes delivered by the male owl in the general proximity of a potential nesting site. As the breeding season progresses and the nest site is occupied by the female, the song diminishes in its frequency.

The prolonged song, also given only by the male, is a softer and longer version of the primary song. The male will produce this call as he flies back and forth between the female and the proposed nesting cavity. The song continues throughout the courtship and up to the time that the eggs begin to be incubated. It seems to function in the pair-bonding process. The male produces a shorter version of this song as a delivery vocalization when approaching the nest with food for the incubating female or recently hatched young.

Among other sounds produced, pairs of boreals will screech to maintain contact with each other in their territory. The female owl produces a soft peeping call particularly when responding to the male's delivery call as he approaches the nest with prey. She also produces a harsh *"chuuk"* call when in the male's territory, and in response to his prolonged song.

As in other owl species, these birds register their irritation over being disturbed or handled by hissing and bill-snapping.

COURTSHIP AND NESTING

In some parts of their range in North America, male boreal owls establish territories and begin their courtship calls as early as late January, although this activity can be delayed as the influence of weather and prey availability demands. The male may sing from several cavities in his effort to attract a mate, but ultimately this monogamous relationship is sealed when the female determines where she will lay her eggs.

The timing of nest occupancy and egg laying by the female varies from region to region, with some birds in Minnesota initiating clutches as early as the end of March while other birds in Idaho start incubation as late as mid-April and into the third week of May. The incubation period of the two to three eggs lasts an average of twenty-nine days, and through the first three weeks of this period the female is dependent on the male for provisions.

Twenty-seven to thirty-two days after hatching, Idaho boreal owls begin fledging from their nest cavities. Once out, the youngsters remain loosely grouped for a week or more in the general vicinity of the nest, where the parents continue to feed them. By their third week, the young owls have begun to wander from their natal site, and feeding by the parent owls has lessened or ceased. At six weeks they are on their own.

THREATS AND CONSERVATION

There are a number of forest predators the boreal owl must be alert to avoid. Pine martens are quick to exploit the vulnerability of nesting boreals, taking both nestlings and brooding females. Goshawks, Cooper's hawks, and great horned, spotted, and barred owls are all predators for this species.

The odds of a bird's surviving its first year after fledging are very low. Avoiding predators is one thing, but locating and capturing prey is a particular challenge for first-year owls. In Europe, Erkki Korpimaki's work with boreal owls found that nearly 78 percent of the fledgling male birds died before their first breeding attempt.

The cutting of forests throughout the range of this species is the greater threat to the bird's welfare. Nesting, foraging, and roosting requirements of boreals are compromised or destroyed as the increasingly fragmented stands of trees are cut. Removal of the older trees suitable for woodpecker-dug cavities is an obvious detriment to this cavity-dependent owl. Plantation forestry with its evenly aged trees does not allow for the older snags to remain, and with their absence the owls have no nesting options. There is a promising alternative, however. In some parts of this owl's range it has occupied nesting boxes when proper woodpecker cavities are unavailable.

As with all owls, the role of conservation education is important. What might be overlooked or treated indifferently in matters of forest management, but have a telling effect on the birds, can be addressed directly with training of personnel in natural resources agencies on the subject of native owls and their habitat requirements. A simple public information brochure related to the dynamics of local forests and their resident owls will elevate both understanding of the species and the valuing of them. In biology classrooms, using owls as examples when discussing regional ecology not only clarifies the subject but also makes students aware of these species and their significance.

VITAL STATISTICS

Length: 8.5–12 inches (22–30 centimeters)
Wingspan: 19–25 inches (48–63 centimeters)
Weight: 3.5–5 ounces (100–140 grams)

6.3 The full head of a great gray owl,
showing the huge feathered disks around its eyes.

Great Gray Owl (*Strix nebulosa*)

For a naturalist, first encounters of particularly impressive species tend to be indelibly etched in memory. Samuel Rathbun, an important ornithologist in Washington state in the early part of the twentieth century, was fond of recounting his unusual first sighting of the great gray owl. He had been out strolling along a sidewalk in downtown Seattle when a young girl pushing a baby carriage approached him from the opposite direction. They passed each other, and he continued to walk a few more steps before he abruptly stopped in his tracks, realizing that there was something unusual about the baby he'd seen when he looked down into the carriage. Turning around, he caught up with the little girl and, checking the contents of the carriage, gazed down on the broad round face of a great gray owl fully attired in doll clothes, bonnet and all.

My first meeting of this immense owl, while not as unique as Rathbun's, was a good deal livelier and equally serendipitous. Friends and I had gone into the Cascades north of the Skagit River pursuing a vague rumor that one of these birds had been seen around Baker Lake, in western Washington. If one sets out to find a species, I think it is of some value to think a bit like the creature you pursue. Knowing a little bit about the bird's preference for open meadows between forest stands, we stopped at the first one we came to, and there indeed was the owl.

Because of its size and the light gray of its plumage, this species stood out against the dark backdrop of the winter landscape. More than three hundred feet (a hundred meters) across the meadow, the owl bobbed its head from side to side, adjusting the resolution of whatever it was watching and listening to. I imagined that it was hunting the meadow voles that dug their shallow tunnels under the ground cover here. Giving all its attention to hunting, the owl changed perches a few times and unconcernedly moved closer to its audience. We had been watching only for a quarter hour at most when this first encounter with the great gray owl suddenly exceeded anything I might have imagined. Perched on top of a small fir less than 160 feet away (50 meters), the owl suddenly raised its wings, pitched off its perch, and flew directly toward me. Reflexively I raised my arm, as I've done to permit a hawk or falcon to land on a glove, but that was not the owl's intent. The owl's flight took it to just above my head, and as it passed I felt the soft brush of its feathered toes over the back of my hand.

RANGE AND HABITAT

The great gray occupies the North American and European continents, favoring the latitudes of the taiga and boreal forests. In North America some of this species has sustained isolated breeding populations southward from the far north along the Cascade and Rocky Mountain ranges to occupy the more open locations of the bogs and meadows in the Yosemite and Yellowstone National Parks.

FOOD PREFERENCES

Remarkably well insulated, this owl can remain in its northern habitats through the severest winters and is able to take prey beneath the snow cover. Although voles make up the bulk of its diet, it captures a range of other small mammals, including deer mice, jumping mice, and grasshopper mice. The great gray, in spite of weighing 15 percent less than the great horned owl, can capture large mammals, including snowshoe hares and short-tailed weasels. Occasionally they eat birds, including jays, robins, and grouse.

VOCALIZATIONS

There are several basic calls of the great gray that include a "hooot—Hoot," employed by the male as an advertising call in its territory; a "whooop—whoop," the

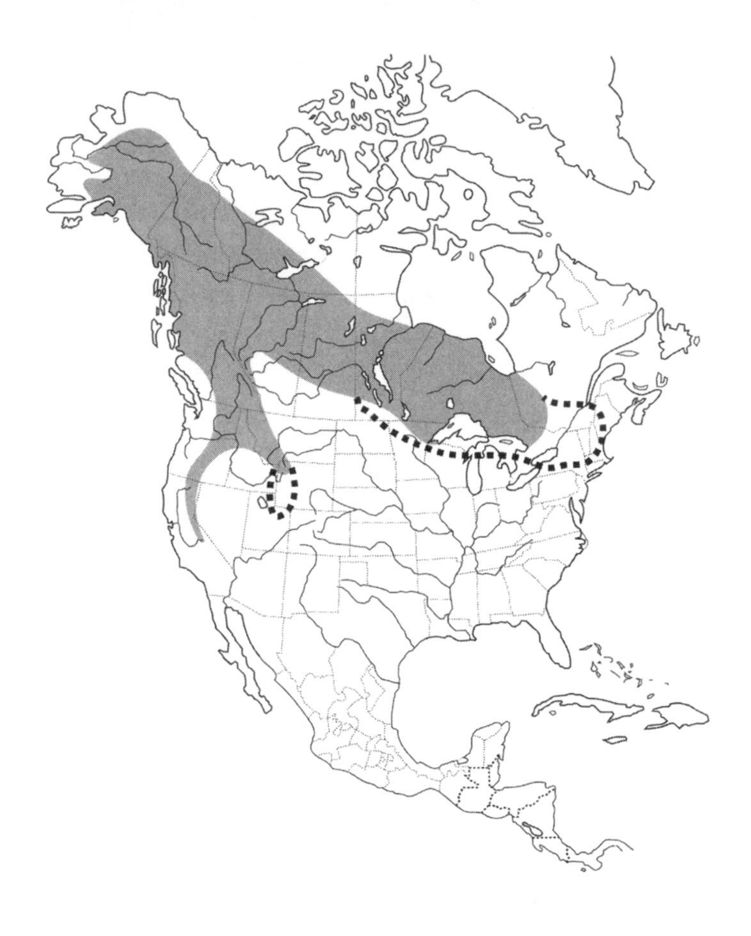

6.4 Range map of great gray owl in North America.
(Dashed lines show the extent of occasional occurrence of the species.)

food solicitation call of the female; hisses uttered by the nestlings perhaps as a food soliciting vocalization; chitters delivered between adults when food is exchanged; and bill snaps expressed when the owl is irritated or threatened.

COURTSHIP AND NESTING

Like other owls, the great gray is monogamous and by the second year will begin breeding. Early in the year, January and February, the male is displaying to the female and demonstrating his prowess as a hunter by plunging into the snow, and if successful the catch is delivered to his prospective mate. The female will inspect the possible nesting sites the male may introduce her to, but she alone will make the final choice. In these early stages of courtship, to strengthen the pair bond, there is much mutual preening between two birds.

Depending on the location and the severity of the winter, three to five eggs may be laid as early as March or be delayed until early May. In North America the incubation period averages nearly thirty days, and throughout this time and the early brooding of the young, the male is the sole provider of food. Should the young be threatened, both adults are vigorous defenders of their family. Within a month after hatching the young are jumping and fluttering from the nest.

Although out of the nest, the owlets will require at least another week before they can fly. In spite of this ability, they can, as do other owl species, use their feet and flapping wings to scale fallen trees and low limbs, allowing them to roost off the ground. The adults remain close at hand and for as long as another three months they will continue to feed their young. Thereafter, the young owls will begin dispersing from their natal sites.

THREATS AND CONSERVATION

In spite of the adult owls vigorously defending their nestlings, goshawks, great horned owls, and red-tailed hawks are serious threats to the young. The more exposed nests are open to the predations of ravens.

Not only the predators but also the possibility of starvation is ahead for the recently independent birds. One study found that the owl had little more than a fifty-fifty chance of surviving its first year. Moving to new territory can expose the birds to the human-impacted environment where collisions with cars, wire fences, and electrocution are possible.

Altered landscapes are always threats to owls, particularly when a forest is cut to remove a majority of the larger trees and snags, reducing the availability of roosting and nesting sites along with locations where they can perch and scout for prey.

It is a good sign that in parts of their range the owls readily take to nesting on wooden platforms or wire baskets covered with a stick substrate and placed in locations where alternative nesting sites are unavailable. An artificial nest site at least sixteen feet (five meters) high and inside a forest will provide additional protection and shading for the birds. Forest management practices that selectively open up contiguous stands of second-growth trees will encourage the presence of small mammals that will soon become a source of food for the great grays.

6.5 Hawk owl pursuing a gray jay.

VITAL STATISTICS

The great gray owl appears to be capable of a long life, with one wild owl banded and recovered after thirteen years.

Length: 24–33 inches (61–84 centimeters)
Wingspan: 54–60 inches (135–150 centimeters)
Weight: 1.5–3 pounds (680–1,360 grams)

Northern Hawk Owl (*Surnia ulula*)

One summer I joined a group of artists from other countries to respond artistically to the vast wilderness that composes the Copper River delta in the south-central region of Alaska. "You could put the entire country of Holland here," exclaimed a painter from the Netherlands as he swept his arm while pointing to the seven hundred thousand acres of river delta wedged between the Chugach and Wrangle Mountain ranges. It was here that two friends from England and I set out to

6.6 Range map of hawk owl in North America.

see if we might find a northern hawk owl that had recently been spotted. Given the thousands of square miles that constitute this area, our quest to find, study, and paint this single owl seemed daunting.

As we were driving along one stretch of the road paralleling the river, the three of us discussed the particular characteristics that set this species apart from so many others of its family. It is a "diurnal owl, built more like a big-headed Cooper's hawk than an owl," opined my friend Andrew Haslen, from Suffolk. "Aye, and perches on prominent snags much as a hunting buteo would," offered the painter David Bennett from London. It was shortly after this last description had been offered that I looked up from the road toward the trees ahead and exclaimed: "There's one right there!"

The owl, perched high in a cottonwood just off the road, seemed to hardly notice us as we frantically pulled our gear from the back seat, and, with the help of binoculars, spotting scopes, and sketch pads, set to work to record our impressions of this distinguished owl. Even as we scurried about, the northern hawk owl remained relaxed on its perch and gave us only an occasional glance as it scanned the flat landscape beyond the trees. We got very good views, but they were all from one side. Determined to get a frontal view, I took camera and binoculars in hand and started to wade into the side channel that fronted the cottonwood stand. Although my first step was into the shallows, my next step took me into the depths. In a second I was up to my neck in the glacially fed waters of the Copper River with my hands aloft holding my gear.

Suffice it to say that my artist chums had one of the better laughs of the expedition at my expense as I floundered about trying to get a footing on a slippery river bottom to throw myself back onto the bank. I remember feeling the divots under my sodden shoes that had been left earlier in the day by the hooves of willow-browsing moose. Through it all the northern hawk owl remained overhead, finding very little of interest in the floundering below.

RANGE AND HABITAT

A diurnal owl, it occupies the northern forests across the North American continent and into northern Europe and Asia. In winters after an abundant rate of reproduction, if prey is scarce, hungry northern hawk owls may be seen as far south as the states of Washington, Nebraska, and Illinois. Recent reports have these owls breeding in the northeastern portions of Washington state.

FOOD PREFERENCES

This medium-sized owl, physically evolved to be something of a match for a Cooper's hawk, is a versatile diurnal hunter. Lacking the asymmetrical position of ear openings that allows nocturnal owls to locate prey virtually unseen, their keen eyesight permits the sighting of prey as far away as 900 yards (800 meters).

Cutting a sleek, swift profile in their swooping flights, they pursue and capture voles and lemmings as a mainstay of their diets. Hawk owls will also take moles, as well as larger prey such as rabbits, snowshoe hares, and red squirrels. Birds caught

6.7 Hawk owl attacking a spruce grouse.

include starlings, sparrows, robins, jays, and even the larger pileated woodpecker and spruce grouse.

VOCALIZATIONS

This owl possesses a wide array of vocalizations; what follows are a few of them.

An "ululululululul" is produced as a trill whistle sequence by the male to advertise his territory. It may be delivered when the bird is both perched and in flight display.

The "rike, rike, rike, rike" cry sequence is given by both male and female to protest other northern hawk owls entering their territory, but also as an approach call when the male brings food to the nest and his brooding mate.

"Kiiiiiiue" is a sharp trilling call used by the owl if the nest is disturbed or a threatening intruder approaches the nest and young.

The "chut, chut" call is given by the female as a sharp vocalization during copulation.

COURTSHIP AND NESTING

The males of this monogamous owl are issuing their territory advertising and lure calls to the females very early in the New Year. By February the duets of pairs of northern hawk owls can be occasionally heard, and in North America these birds can be laying eggs as early as mid-March and will continue to as late as June.

The female exclusively incubates a clutch of eggs for twenty-five to twenty-nine days. She typically lays them in the top of a broken snag or pileated woodpecker cavity. As with other species of owls, supplying food to the female and then to the hatchlings falls to the male. These birds are very attendant to their nest and visit the site three to four times more than other northern owls.

By the third or fourth week from hatching, the young owls are prepared to jump, flutter, or climb down from the nest. Throughout the breeding period the adult owls are vigorous defenders of their family, particularly after the young have hatched and fledged. Defending against predatory birds as large as goshawks, and inquisitive humans, the northern hawk owls stoop and scream at interlopers, seeking to drive them from the vicinity of the nest. Female owls have been observed, not unlike the long-eared owl, to fly to the ground and feign injury in an effort to distract a predator from her brood.

In the European portion of the northern hawk owl's range, the young owls are independent by their third month out of the nest and separate entirely from their parents and the natal site.

THREATS AND CONSERVATION

Predators occupying the northern hawk owl's habitat include the goshawk, great horned owl, fisher, and pine marten. In more open areas, peregrine falcons and gyrfalcons sometimes capture these birds.

A greater threat continues to be the occasional shooting of this species. Innocent of human intent, the bird is easily approached. It perches prominently where it can scan for prey, and because of this habit presents an obvious silhouette to the impetuous and irresponsible outdoorsperson interested in shooting. In parts of its Canadian range, this bird is referred to as a "target owl."

Clear-cuts through pristine northern forests reduce the availability of nesting sites for this species, and although such open areas may attract vole populations, the absence of suitable trees for raising young, observing, hunting, and roosting will affect their numbers. Taking mitigating measures when cutting can reduce these impacts, such as leaving snags and cutting selectively for a mix of open areas for prey and sufficient trees to allow the owl breeding and observation locations. In parts of its European range, where the snag availability is scant, nest boxes have proven valuable. In general, however, respecting its existing ecological requirements appears to be the most efficient way to maintain what currently appears to be a healthy population of this dashing species.

VITAL STATISTICS

In captivity, the northern hawk owl has lived to ten years.

Length: 14.5–17.5 inches (36–44 centimeters)
Wingspan: 31–35 inches (80–90 centimeters)
Weight: 12–14 ounces (340–396 grams)

Snowy Owl (*Bubo scandiacus*)

Where I live outside Seattle it is not unusual every year or two to discover a few snowy owls taking up winter residency. In peak irruption years the owls are numerous enough to function as a public relations firm for their entire tribe.

Catching glimpses of a small cluster of snowy owls perched like diminutive snowmen on drift logs is among my greatest pleasures as a naturalist and artist. At the same time, other more intimate encounters have provided me with the informa-

6.8 Snowy owl flying over slough.

tion and inspiration to interpret the birds in line and carved stone. Because these wintering owls are most often from the Arctic wilderness, and hatched earlier the same year, they have no experience with the dangers of urban settings, and many are beset by injuries resulting from collisions with vehicles, windows, and wires. Injured birds came my way from time to time, and in seeking to help them recover I was able to study them more completely.

When I took one owl from the blanket it was wrapped in, I recall feeling the sharpness of its breast bone and knew that it was underfed. At the same time, its strength and defiant expression indicated just the opposite. In spite of my cautiously handling it with thick leather gloves, the bird's foot flashed out to grab my thumb with its toes and sink a talon into it. Getting the owl to relinquish its grip was not accomplished with an attempt to pry off its toes. Such is the strength of the bird that only by waiting for the snowy to calm down and relax its grip was I able to withdraw my hand.

Although momentarily painful to me and frightening to the owl, handling the bird provided a clearer picture of its condition and its physical composition. No broken bones, but certainly underweight and needing a rich and sustained diet to regain its full vitality. By holding the owl's legs together with my gloved hand and tracing the bird's form with the fingers of my other hand, I could begin to register the information that would help me interpret the bird artistically. Its muscular upper legs hidden by flank feathers certainly accounted for the fullness of its lower body. The heavy musculature on either side of its sternum accounted for its broad chest and suggested how such a large land bird could fly the thousands of miles southward to winter here.

Rather than imagining what its plumage might be like from watching the snowy owl at a distance, here I could see and feel it. My sense of touch came in contact with the distinctive silky strands of feathering that seem to flow so thickly over the bird's foot and toes. I felt the many layers of feathers over the chest and flanks that combine to form several inches of warm air-trapping insulation to assure the bird's survival in severe cold. I considered the patterns of feather tracks across the nape of the owl's neck and those of the shoulders, wherein the upper portions of the bird's great wings can be tucked when at rest.

Once, while examining a dead snowy, I noticed distinct white flecks of debris on the outer edges of the bristles that extended over its beak. On closer inspection I was astounded to discover that the individual specks were moving. The owl was infested with feather lice, Mallophaga. The color of the lice was a precise match for the male owl's white plumage. Their movement forward to the owl's face upon its death made it possible for me to discover them.

Like all birds, owls are subject to external and internal parasites, but this one surprised me. It had coevolved with the owl to match the bird's color and become nearly invisible. A farsighted owl would have a hard time removing them. I discovered another interesting phenomenon when I placed the dead owl in the freezer for a week to keep it fresh before skinning it. When I removed the rock-solid frozen owl, there were still live lice moving slowly about on the face of the bird—a suggestion that these parasites had developed an internal antifreeze that would give them some additional assistance in extending their lives.

RANGE AND HABITAT

The snowy owl is a diurnal circumpolar species of owl residing and breeding beyond the tree line on the open tundra in the Arctic of both North America and northern Europe. As discussed previously, some of these birds periodically move south in some winters through Canada to reach the northern portions of the contiguous United States.

6.9 Range map of snowy owl in North America.

FOOD PREFERENCES

These opportunistic feeders can make do when lemmings are scarce, and shift their tastes to other mammals, particularly voles. When wintering outside their normal range they capably capture rabbits and ground squirrels and are quick to seize birds as well, including assorted passerines, grebes, and grouse. Adaptable, one wintering snowy owl took up residency near my home to remain nearly a month as it feasted on a colony of mountain beaver.

The food energy required by this owl is substantial. It is estimated that an active male snowy owl requires nearly a pound (more than four hundred grams) of food a day, which adds up to about a quarter of its body weight.

VOCALIZATIONS

A "hoo-hoo" is delivered fortissimo in a sequence of up to six calls, and serves as a territorial advertisement by the male. In the Arctic, it can be heard over a distance of several miles.

Their "ha-how, quack-guack" series of calls is produced on the ground or in flight and is associated with the birds confronting a threat.

A "ka-ka-ka—ka" is a call particular to a male's response to a threat, at which time the bird may also produce a rattling and a "rick-rick—rick" sound.

The "ke-ke-ke-ke-ke" series of calls is associated with both coition and the feeding of the young owls.

COURTSHIP AND NESTING

The male owl is quite demonstrative in his courtship display, taking to the air in undulating flight with exaggerated wing beats that carry the bird past his mate, and ending in a gradual vertical descent to the ground. Often a lemming is carried in the beak or feet as a further enticement to the female. A ground display follows, with wings partially out as the male obliquely faces the female, head lowered and body stretched out. When both birds assume this posture, coition often follows.

These Arctic owls are late nesters, waiting until the snow cover on the tundra begins to diminish in mid-April into mid-May. Selecting a location at higher elevations helps in providing drainage and a good view of the surrounding landscape, and the birds will also seek to keep the nest close to a dependable food supply. They form ground nests that are essentially bowl-like scrapes on the surface, and they appear to locate the nest in such a way that it is windswept, keeping this area dry and snow free.

The female may lay as many as eleven eggs and is solely responsible for their incubation and the brooding of the young. Such a number of eggs means a significant physical investment by the female owl, representing up to 43 percent of the bird's body weight.

The young begin to hatch at different intervals, with the first owlet, from the first egg laid, emerging at about thirty-one days. Within two weeks of hatching the owls are capable of shuffling out from the nest site on foot, and by the third week they are wandering about in its general vicinity.

For another five weeks after the owlets have exited the nest, the parents continue to feed the youngsters—a labor in itself, for by the time of their independence it has been estimated that a family of nine owlets will have consumed a total of fifteen hundred lemmings.

THREATS AND CONSERVATION

Exposed as they are as a ground-nesting species, snowy owls (particularly the young) are subject to predation by Arctic foxes, jaegers, and occasionally a gyrfalcon. Starvation is always a possibility as the prey base ebbs and flows over the years.

Fatalities among snowy owls are often large during their irruptive periods when they venture into new territories south of their normal remote range, as they may collide with vehicles and utility lines and be electrocuted or even shot.

It appears that populations of snowy owls wax and wane depending largely on their food supplies through breeding and wintering seasons. In the Canadian Arctic a survey taken in the 1950s on Banks Island estimated that the number of owls ranged from fifteen thousand to twenty thousand during a high rate of reproduction. Another measure in the same area during a period of low reproduction revealed only two thousand birds.

A future that includes snowy owls is somewhat in doubt. With the conditions of global warming, the dynamics of the Arctic ecosystem may be altered to the point where this owl as a wild species will struggle. Although successful captive reproduction of snowy owls has been achieved, there is no assurance that long term there will be a suitable environment where these birds may be returned.

In the meantime, federal laws protecting all owls must be vigorously enforced and public education pursued. An informed and engaged public will respect the species and behave in a way that stewards its presence when wintering in the proximity of human settlements. Keeping a critical distance of more than three hundred feet (one hundred meters) from birds while observing them will give them the space and solitude required to hunt, rest, and begin the initial pair bonding that will assure their return to produce new generations in their Arctic environment.

VITAL STATISTICS

A captive snowy owl lived for twenty-eight years in Switzerland, and a wild bird is known to have reached the age of nine and a half years.

Length: 20–27 inches (51–68 centimeters)
Wingspan: 54–66 inches (140–170 centimeters)
Weight: 3.5–4.5 pounds (1.5–2 kilograms)

Bibliography

Angell, Tony. *Owls*. Seattle: University of Washington Press, 1974.

Angell, Tony, and John Marzluff. *Gifts of the Crow.* New York: Simon & Schuster, 2012.

Angell, Tony, and John Marzluff. *In the Company of Crows and Ravens.* New Haven: Yale University Press, 2005.

Backhouse, Francis. *Owls of North America.* Buffalo: Firefly Books, 2008.

Baldini, Umberto. *The Sculpture of Michelangelo.* New York: Rizzoli International Publications, 1981.

Bannick, Paul. *The Owl and the Woodpecker.* Seattle: Mountaineers Books, 2008.

Barnard, Jeff. "Feds Advance Plan to Kill Barred Owls in Northwest." *Seattle Times,* July 23, 2013.

Bates, Marston. *The Forest and the Sea: A Look at the Economy of Nature and the Ecology of Man.* New York: Vintage Books, 1960.

Bent, Arthur Cleveland. *Life Histories of North American Birds of Prey,* part 2, vol. 170, United States National Museum Bulletin. Washington, D.C.: Smithsonian Institution, 1938.

Borror, Donald Joyce, and Richard E. White. *A Field Guide to the Insects of America North of Mexico.* Boston: Houghton Mifflin, 1970.

Bowers, Nora and Rick. *Mammals of North America.* New York: Houghton Mifflin, 2004.

Bull, E. L., and J. R. Duncan. 1993. Great Gray Owl (*Strix nebulosa*). In *The Birds of North America,* no. 41 (A. Poole and F. Gill, eds.). Philadelphia: The Academy of Natural Sciences; Washington, D.C.: The American Ornithologists' Union.

Burton, John A., ed. *Owls of the World.* New York: Eurobook, 1973.

Cannings, R. J. 1993. Northern Saw-whet Owl (*Aegolius acadicus*). In *The Birds of North America,* no. 42 (A. Poole and F. Gill, eds.). Philadelphia: The Academy of Natural Sciences; Washington, D.C.: The American Ornithologists' Union.

Cannings, R. J., and T. Angell. 2001. Western Screech Owl (*Megascops kennicottii*). In *The Birds of North America,* no. 597 (A. Poole and F. Gill, eds.). Philadelphia: The Academy of Natural Sciences; Washington, D.C.: The American Ornithologists' Union.

Cruickshank, Helen. *Thoreau on Birds.* New York: McGraw-Hill Book Company, 1964.

Duncan, James R. *Owls of the World.* Buffalo: Firefly Books, 2003.

Duncan, J. R., and P. A. Duncan. 1998. Northern Hawk Owl (*Surnia ulula*). In *The Birds of North America,* no. 356 (A. Poole and F. Gill, eds.). Philadelphia: The Academy of Natural Sciences; Washington, D.C.: The American Ornithologists' Union.

Elliot, George. *Sculpture of the Inuit: Masterworks of the Canadian Arctic.* Toronto: University of Toronto Press, 1971.

Everett, Michael. *A Natural History of Owls.* London: Hamlin Publishing Group, 1977.

Fox-Davies, Arthur C. *Heraldry: A Pictorial Archive for Artists and Designers.* Mineola: Dover Publications, 1991.

Gehlbach, Frederick. "Eastern Screech Owl Responses to Suburban Sprawl, Warmer Climate, and Additional Avian Food in Central Texas," *Wilson Journal of Ornithology,* no. 3 (2012): 631–634.

Gehlbach, Frederick. "Body Size Variation and Evolutionary Ecology of Eastern and Western Screech Owls," *Southwestern Naturalist,* no. 48 (2003): 70–80.

Gehlbach, Frederick. "Eastern Screech Owls in Suburbia: A Model of Raptor Urbanization." In *Raptors in Human Landscapes* (David M. Bird, Daniel E. Varland, and Juan Jose Negro, eds.), 69–74. San Diego: Academic Press, 1996.

Gehlbach, Frederick. 1995. Eastern Screech Owl (*Megascops asio*). In *The Birds of North America,* no. 165 (A. Poole and F. Gill, eds.). Philadelphia: The Academy of Natural Sciences; Washington, D.C.: The American Ornithologists' Union.

Gehlbach, Frederick. *The Eastern Screech Owl.* Dallas: Texas A & M University Press, 1994.

Gehlbach, Frederick, and R. S. Baldridge. "Live Blind Snakes in Eastern Screech Owl Nests: A Novel Commensalism," *Oecologia* (1987): 560–563.

Gehlbach, Frederick, and N. Y. Gehlbach. 2000. Whiskered Screech Owl (*Megascops trichopsis*). In *The Birds of North America*, no. 507 (A. Poole and F. Gill, eds.). Philadelphia: The Academy of Natural Sciences; Washington, D.C.: The American Ornithologists' Union.

Gehlbach, Frederick, and Jill Leverett. "Mobbing of Eastern Screech Owls," *The Condor* 97, no. 3 (1995): 831–834.

Grossman, Mary Louise, and John Hamlet. *Birds of Prey of the World.* New York: Clarkson N. Potter, 1964.

Guitérres, R. J., A. B. Franklin, and W. S. Lahaye. 1995. Spotted Owl (*Strix occidentalis*). In *The Birds of North America*, no. 179 (A. Poole and F. Gill, eds.). Philadelphia: The Academy of Natural Sciences; Washington, D.C.: The American Ornithologists' Union.

Hauge, E. A., B. A. Millsap, and M. S. Martell. 1993. Burrowing Owl (*Speotyto cunicularia*). In *The Birds of North America*, no. 61 (A. Poole and F. Gill, eds.). Philadelphia: The Academy of Natural Sciences; Washington, D.C.: The American Ornithologists' Union.

Hayward, G. D., and P. H. Hayward. 1993. Boreal Owl (*Aegolius funereus*). In *The Birds of North America*, no. 63 (A. Poole and F. Gill, eds.). Philadelphia: The Academy of Natural Sciences; Washington, D.C.: The American Ornithologists' Union.

Heinrich, Bernd. *One Man's Owl.* Princeton: Princeton University Press, 1987.

Henry, S. G., and F. R. Gehlbach. 1999. Elf Owl (*Micrathene whitneyi*). In *The Birds of North America*, no. 413 (A. Poole and F. Gill, eds.). Philadelphia: The Academy of Natural Sciences; Washington, D.C.: The American Ornithologists' Union.

Holt, D. W., and S. M. Leasure. 1993. Short-eared Owl (*Asio flammeus*). In *The Birds of North America*, no. 62 (A. Poole and F. Gill, eds.). Philadelphia: The Academy of Natural Sciences; Washington, D.C.: The American Ornithologists' Union.

Holt, D. W., and J. L. Petersen. 2000. Northern Pygmy Owl (*Glaucidium gnoma*). In *The Birds of North America*, no. 41 (A. Poole and F. Gill, eds.). Philadelphia: The Academy of Natural Sciences; Washington, D.C.: The American Ornithologists' Union.

Houston, C. S., D. G. Smith, and C. Rohner. 1998. Great Horned Owl (*Bubo virginianus*). In *The Birds of North America,* no. 41 (A. Poole and F. Gill, eds.). Philadelphia: The Academy of Natural Sciences; Washington, D.C.: The American Ornithologists' Union.

Hume, Rob. *Owls of the World.* Philadelphia: Running Press Book Publications, 1991.

Johnsgard, Paul. *North American Owls.* New York: Smithsonian Institution, 1988.

Leonard, Pat. "A Season of Snowy Owls." In *Living Bird* 33, no. 2 (Spring 2014).

MacKinnon, A. *Plants of the Pacific Northwest Coast: Washington, Oregon, British Columbia, and Alaska.* Rev. ed. Vancouver: Lone Pine, 2004.

Marks, J. S., D. L. Evans, and D. W. Holt. 1994. Long-eared Owl (*Asio otus*). In *The Birds of North America,* no. 133 (A. Poole and F. Gill, eds.). Philadelphia: The Academy of Natural Sciences; Washington, D.C.: The American Ornithologists' Union.

Marshall, Joe T., Jr. *Parallel Variation in North and Middle American Screech Owls.* In *Monographs of the Western Foundation of Vertebrate Zoology,* vol. 1. (Jack C. von Bloeker, Jr., ed.). Los Angeles: Western Foundation of Vertebrate Zoology, 1967.

Marti, C. D. 1992. Barn Owl (*Tyto alba*). In *The Birds of North America,* no. 1 (A. Poole, P. Stettenheim, and F. Gill, eds.). Philadelphia: The Academy of Natural Sciences; Washington, D.C.: The American Ornithologists' Union.

Martin, Graham. *Birds by Night.* London: T. & A. D. Poyser, 1990.

Mayor, A. Hyatt. *A Century of American Sculpture.* New York: Abbeville Press, 1981.

Mazer, K. M., and P. C. James. 2000. Barred Owl (*Strix varia*). In *The Birds of North America,* no. 508 (A. Poole and F. Gill, eds.). Philadelphia: The Academy of Natural Sciences; Washington, D.C.: The American Ornithologists' Union.

McCallum, D. A. 1994. Flammulated Owl (*Otus flammeolus*). In *The Birds of North America,* no. 93 (A. Poole and F. Gill, eds.). Philadelphia: The Academy of Natural Sciences; Washington, D.C.: The American Ornithologists' Union.

Mearns, Barbara, and Richard Mearns. *Audubon to Xantus.* San Diego: Academic Press, 1992.

Mikkola, Heimo, and Ian Willis. *Owls of Europe.* Staffordshire: T. & A. D. Poyser, 1983.

Minor, William F., Maureen Minor, and Michael F. Ingraldi. "Nesting of Red-Tailed Hawks and Great Horned Owls in a Central New York Urban/Suburban Area," *Journal of Field Ornithology* 64, no. 4 (Autumn 1993): 433–439.

Morris, Desmond. *Owl.* (Animal series; Jonathan Burt, ed.) London: Reaktion Books, 2009.

Nero, Robert W. *The Great Gray Owl: Phantom of the Northern Forest.* Washington, D.C.: Smithsonian Institution Press, 1980.

Parmelee, David. 1992. Snowy Owl (*Bubo scandiacus*). In *The Birds of North America,* no. 10 (A. Poole, P. Stettenheim, and F. Gill, eds.). Philadelphia: The Academy of Natural Sciences; Washington, D.C.: The American Ornithologists' Union.

Parr, Michael. "Long-Whiskered Owlet: The Bird I Had to See," *Bird Conservation,* Winter 2013–2014, p. 26.

Peeters, Hans. *California Natural History Guides: Field Guide to Owls,* vol. 93. (Phyllis M. Faber, Bruce M. Pavlik, eds.). Berkeley: University of California Press, 2007.

Proudfoot, G. A., and R. R. Johnson. 2000. Ferruginous Pygmy Owl (*Glaucidium brasilianum*). In *The Birds of North America,* no. 498 (A. Poole and F. Gill, eds.). Philadelphia: The Academy of Natural Sciences; Washington, D.C.: The American Ornithologists' Union.

Pyle, Robert Michael. *The Butterflies of Cascadia: A Field Guide to All the Species of Washington, Oregon, and Surrounding Territories.* Seattle: Seattle Audubon Society, 2002.

Rashid, Scott. *Small Mountain Owls.* Atglen, Pennsylvania: Schiffer Publishing, 2009.

Ray, Dorothy Jean. *Eskimo Art: Tradition and Innovation in North Alaska.* Seattle: University of Washington Press, 1977.

Schwartz, John. "A Snowy Owl Influx Thrills, Baffles Birders." *Seattle Times,* February 1, 2014.

Schwartz, John. "A Bird Flies South, and It's News." *New York Times,* January 31, 2014.

Singer, Robert T. *Edo Art in Japan, 1615–1868.* Washington, D.C.: National Gallery of Art, 1998.

Sparks, John, and Tony Soper. *Owls.* New York: Taplinger Publishing, 1970.

Taylor, Marianne. *Owls.* Ithaca: Cornell University Press, 2012.

Tripp, Tiana M. *Use of Bioacoustics for Population Monitoring in the Western Screech Owl (Megascops kennicottii)*. Master's thesis, University of Victoria, 1995.

VanCamp, Laurel. *North American Fauna: The Screech Owl,* vol. 71. U.S. Fish and Wildlife Service, 1975.

Voous, Karel H. *Owls of the Northern Hemisphere.* London: William Collins Sons, 1989.

Walker, Lewis Wayne. *The Book of Owls.* New York: Alfred A. Knopf, 1974.

Warren, Lynne. "Muscle and Magic: Snowy Owls," *National Geographic,* December 2002.

Wheye, Darryl, and Donald Kennedy. *Humans, Nature, and Birds.* New Haven: Yale University Press, 2008.

Whiting, Jeffrey. *Jeffrey Whiting's Owls of North America.* (Whiting's Reference of Birds, vol. 1.) Clayton, Ontario: Heliconia Press, 1972.

Wolfe, Art. *Owls: Their Life and Behavior.* New York: Crown Publishers, 1990.

World Owl Trust, "Owl Information" (www.owls.org). Accessed June 20, 2013.

Yolen, Jane. *Owl Moon.* Illustrated by John Schoenherr. New York: Philomel Books, 1987.

Illustration Credits

Courtesy of Cornell Lab of Ornithology: Figs. 4.2, 4.4, 4.6, 4.8, 4.10, 4.12, 4.14, 4.16, 4.18; 5.2, 5.4, 5.6, 5.8, 5.10, 5.12, 5.14; 6.2, 6.4, 6.6, 6.9

Index